Pablo Pedregal

Introduction to Optimization

Springer

Pablo Pedregal
ETSI Industriales
Universidad de Castilla-La Mancha
13071 Ciudad Real
Spain
pablo.pedregal@uclm.es

Series Editors

J.E. Marsden
Control and Dynamical Systems, 107-81
California Institute of Technology
Pasadena, CA 91125
USA
marsden@cds.caltech.edu

L. Sirovich
Division of Applied Mathematics
Brown University
Providence, RI 02912
USA
chico@camelot.mssm.edu

S.S. Antman
Department of Mathematics
and
Institute for Physical Science and Technology
University of Maryland
College Park, MD 20742-4015
USA
ssa@math.umd.edu

Mathematics Subject Classification (2000): 49-01, 49L20, 90C90, 65K10

Library of Congress Cataloging-in-Publication Data
Pedregal, Pablo, 1963–
    Introduction to optimization/Pablo Pedregal.
       p.cm.—(Texts in applied mathematics; 46)
    Includes bibliographical references and index.
    ISBN 0-387-40398-1 (acid-free paper)
      1. Mathematical optimization. I. Title. II. Series.
    QA402.5.P4 2003
    519.3—dc21   2003053895

ISBN 978-0-387-40398-4
ISBN 0-387-40398-1   Printed on acid-free paper

Printed in the United States of America.

9 8 7 6 5 4 3 2

springer.com

To Daniel,
Silvia,
Jaime,
and Nuria

Series Preface

Mathematics is playing an ever more important role in the physical and biological sciences, provoking a blurring of boundaries between scientific disciplines and a resurgence of interest in the modern as well as the classical techniques of applied mathematics. This renewal of interest, both in research and teaching, has led to the establishment of the series Texts in Applied Mathematics (TAM).

The development of new courses is a natural consequence of a high level of excitement on the research frontier as newer techniques, such as numerical and symbolic computer systems, dynamical systems, and chaos, mix with and reinforce the traditional methods of applied mathematics. Thus, the purpose of this textbook series is to meet the current and future needs of these advances and to encourage the teaching of new courses.

TAM will publish textbooks suitable for use in advanced undergraduate and beginning graduate courses, and will complement the Applied Mathematical Sciences (AMS) series, which will focus on advanced textbooks and research-level monographs.

Pasadena, California   J.E. Marsden
Providence, Rhode Island   L. Sirovich
College Park, Maryland   S.S. Antman

Preface

This book should serve as an undergraduate text to introduce students of science and engineering to the fascinating field of optimization. Several features have been united: conciseness and completeness, brevity and clarity, emphasis on the justification of ideas and techniques and also on applications, etc. One of the novelties of the text is that it ties together fields that are often treated as separate. Indeed, it is hard to find a single textbook where mathematical programming, variational problems, and optimal control problems are explained and integrated as a unity. Thus, our readers may gain an overall view of all aspects of optimization.

It is also true that each of the chapters is but a timid introduction to such broad subjects as linear programming, nonlinear programming, numerical optimization algorithms, variational problems, dynamic programming, and optimal control. As a primer in optimization, our aim with this text is no more than to provide a succinct introduction to those worlds, presented in a single resource reference. This text cannot and does not pretend to substitute in the least other more profound textbooks on those subfields of optimization. Readers with some experience in optimization seeking a more specialized source in some of those parts will have to look for other references. Real-world applications are also far from this introduction to the subject. Although we have tried to motivate the ideas and techniques by using examples, these are most of the time academic simplifications of much more complex situations. Many of our examples and exercises are part of the standard collection of problems often used to introduce optimization. Many of these, even in a much more general form, can also be found in other textbooks.

Applied mathematicians, physicists, and all types of engineers and scientists, may benefit from such an introduction to optimization that does not pay much attention to formalities, technicalities, rigorous proofs, and statements, in order to produce a brief text stressing the main ideas and the main reasons for techniques. We have also tried to keep prerequisites to a minimum. Linear algebra, calculus, and differential equations are essentially the only fields where elementary knowledge is assumed. We hope to help students understand the first principles of optimization so that they may be able to start solving some of the problems they are interested in, and deepen their knowledge of a particular area when needed.

I would like to thank Eduardo Casas, Carlos Corona, Julio Muñoz, and Antonio Ornelas for their reading of the manuscript and for the various, interesting remarks they made. My thanks also go to the staff at Springer, particularly Achi Dosanjh, Joel Ariaratnam, Frank Ganz, Margaret Mitchell, Timothy Taylor, and Elizabeth Young. They all made the preparation of the manuscript and the review process a rewarding and enjoyable task. I am well aware that errors, inaccuracies, ambiguous statements and explanations, misprints, etc., are still part of this text. Anyone interested in letting me know will be welcome to do so by contacting me at *pablo.pedregal@uclm.es*.

Pablo Pedregal
Ciudad Real, April 2003

Contents

Chapter 4. Approximation Techniques

Calculus of Variations

Chapter 5. Variational Problems and Dynamic Programming

Chapter 6. Optimal Control *Theory*

Chapter 1

Introduction

1. SOME EXAMPLES

We believe that there is no better way to convince our readers of the interest and applicability of certain mathematical ideas or techniques than to show the type of practical problems and situations that can be tackled, and eventually solved, by using them. At the same time, this initial list of problems and examples may serve as a clear statement of the objectives and goals of this text. Some of the examples might not be completely understandable in a first reading. This should not bother our readers, since we will insist on them throughout this chapter and their significance will be more clearly grasped by the end of it. Most of the examples we will analyze are very well known and academic, in the sense that the size of real problems is not comparable, in the least, to the situations we will study. More complex versions of these problems can be found in advanced textbooks. We think, however, that the main ideas will be conveyed through them and will endow readers with the basic tools for more realistic situations.

The transportation problem. *A certain product is to be shipped in amounts* $u_1, u_2, \ldots, u_n$ *from n service points to m destinations, where it is to be received in amounts* $v_1, v_2, \ldots, v_m$. *See Figure 1.1. If the cost of sending one unit of product from origin i to destination j is known to be* c_{ij}, *determine the quantity* x_{ij} *to be sent from origin i to destination j so that the total transportation cost is minimum.*

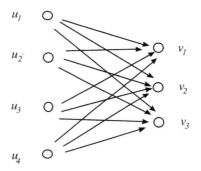

Figure 1.1. A transportation network.

The diet problem. *The nutritive contents of certain foods are known as well as their prices and the daily minimum required for each nutrient. The task consists in determining the amount of each food that must be purchased to ensure that the minimum required for each nutrient is met and the total cost of the diet is as small as possible.*

The scaffolding system. *Consider the scaffolding system of Figure 1.2, where loads* x_1 *and* x_2 *are applied at certain points of beams 2 and 3, respectively. Ropes A and B can bear a maximum weight of 300 kg each, C and D can bear 200 kg, and E and F, a maximum of 100 kg each. Find the maximum load* $x_1 + x_2$ *the system can bear without failure in equilibrium of forces and* moments, the optimal loads x_1 and x_2, and the optimal points where they must be applied, assuming that the weight of ropes and beams is negligible.

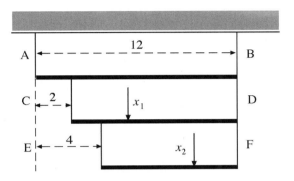

Figure 1.2. Scaffolding system.

Power circuit state estimation. The state variables of an electric network are the voltages, each a complex number with modulus v_i and argument δ_i, at each node of the network. The active and reactive powers of the connection between the nodes i and j are given, respectively, by

$$p_{ij} = \frac{v_i^2}{z_{ij}} \cos \theta_{ij} - \frac{v_i v_j}{z_{ij}} \cos(\theta_{ij} + \delta_i - \delta_j),$$

$$q_{ij} = \frac{v_i^2}{z_{ij}} \sin \theta_{ij} - \frac{v_i v_j}{z_{ij}} \sin(\theta_{ij} + \delta_i - \delta_j),$$

where the modulus z_{ij} and the phase θ_{ij} determine the impedance of the line ij. If experimental measurements $\overline{v}_i, \overline{p}_{ij}, \overline{q}_{ij}$ of the respective values v_i, p_{ij}, and q_{ij} are available, and the parameters of the goodness of the measurements are $k_i^v, k_{ij}^p, k_{ij}^q$, respectively, estimate the state of the network by minimizing, on the variables v_i, the mean quadratic error of the available measurements with respect to the predicted values so that the above formulas hold in the best way possible.

Design of a moving solid. We wish to design a solid with radial symmetry around a given axis that must travel in a straight line with constant velocity

within a fluid. If the density of the fluid is sufficiently small, then the modulus
of the normal pressure in the direction of the outer normal to the surface of
the body exerted by the fluid over the solid comes in the form

$$p = 2\rho v^2 \sin^2 \theta,$$

where ρ and v are the (constant) density and the (constant) velocity of the fluid
relative to the solid, and θ is the angle formed by the tangent to the profile of
the surface in the xy-plane and the velocity of the fluid (see Figure 1.3). How
can we find the optimal profile of the solid in order to minimize the pressure
exerted by the fluid on it?

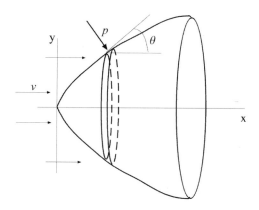

Figure 1.3. A moving solid within a fluid.

Design of a channel.   Channels are a particular type of conducting device
for fluids. Typically, the fluid does not ocupy all of the channel (Figure 1.4),
and in general, losses originate at the walls.

In some specific regime, friction can be approximated by the expression

$$\frac{1}{\sqrt{f}} \approx 2\log\frac{3.7D_h}{e},$$

where f is the friction coeficcient, D_h is the so-called hydraulic diameter, and
e represents a measure of rugosity. Moreover, we have

$$D_h = 4R_h, \quad R_h = A/P,$$

where A is the (area of the) cross section of the channel ocupied by the fluid, and P is the perimeter reached by the same cross section of fluid. If we assume that A is fixed, the question is to determine the profile of the cross section of the channel that will minimize losses of fluid through the walls.

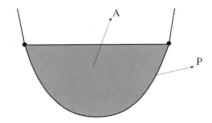

Figure 1.4. The cross section of a channel.

Boat manufacturer.   A boat manufacturer has the following commitments for a certain year: at the end of March, one boat; in April, 2; in May, 5; at the end of June, 3; during July, 2, and 1 in August. He can build a maximum of four boats per month, and can keep three in stock at most. The cost of each boat is 10,000 euros while keeping one in stock is 1,000 euros per month. What is the optimal strategy for building the boats so as to minimize costs?

The harmonic oscillator with friction.   A control surface in a flying object must be kept in equilibrium in a certain position. The fluctuations move the surface, and if they were not addressed, it will vibrate according to the law

$$\theta'' + a\theta' + \omega^2\theta = 0,$$

where θ is the angle measured from the desired equilibrium position, and a and ω are given constants. A servomechanism applies a torque that changes the behavior of the oscillator to

$$\theta'' + a\theta' + \omega^2\theta = u,$$

where the control u must be bounded $|u(t)| \leq C$. The problem consists in determining the servomechanism parameter $u(t)$ such that the surface goes back to rest $\theta = \theta' = 0$ from an arbitrary state $\theta = \theta_0$, $\theta' = \theta'_0$, in minimum time.

A positioning problem. *A certain mobile object moving in the plane is controlled by two parameters: the magnitude of acceleration r and the rate of change of the angle of rotation θ'. If we assume that r and θ' are allowed to move on the intervals $[-a, a]$, $[-\alpha, \alpha]$, respectively, determine the optimal strategy to bring the mobile object from some initial conditions to rest at the origin.*

Although the collection of problems and situations could be considerably enlarged (including some examples, as suggested earlier, closer to reality and to technological or engineering situations), the ones stated above may already serve to suggest that we are before a subject of a relevant applied character. We will be learning to treat and solve these problems and many more in the chapters that follow. Once those ideas have been understood and matured, the reader will be able to analyze and solve by himself (herself) many more situations from science and technology. He (she) may also choose to deepen his (her) knowledge of a particular class of problems by looking for more advanced textbooks on that particular area.

2. THE MATHEMATICAL SETTING

The examples of the previous section are apparently very different among themselves, although they all share something that enables them to be present in this book. In all of those situations we are seeking an optimal solution, the best way to do things, the most efficient manner, the most economical process. Because of this, all of the ideas that have been developed over the years to examine and solve these problems can be put under the label of OPTIMIZATION. Yet, the above problems are very different from one another, and the techniques to solve them or approximate their solutions reflect this same variety and wealth. We do not pretend at this point that readers may discover by themselves these differences, even more so before putting them in a more precise, quantitative fashion reflecting faithfully each situation and allowing an appropriate treatment leading to the solution or a good numerical approximation of it. This process of going from the statement in plain words of a particular situation to its formulation in precise, mathematical terms is of such importance that failure to carry it out accurately may result in absurd answers to problems. The ultimate success of a certain optimization technique greatly depends on it.

The statement of the problem in precise mathematical terms should reflect exactly what we desire to solve. In particular, in dealing with optimization problems there are two important steps to cover. Firstly, the objective or cost function must measure faithfully our idea of optimality. A more desirable solution must have a smaller (or greater) cost functional, be a minimum time, a greater efficiency, greater benefits, minimum losses, etc. If our cost functional does not correctly reflect our optimization criterion, the final solution will not presumably be the optimal situation sought. Secondly, it is equally important to explicitly state the constraints that must be enforced so that admissible solutions are truly feasible in our problem or situation. Once again, if these restrictions are not accurately written, some of them are forgotten, or we are enforcing several that are too restrictive, our final answer may not be what we are looking for. With the aim of emphasizing these issues, we are going to treat, sucessively, the previous problems and provide their mathematical formulation. Before proceeding to such an endeavor, let us indicate some general comments to bear in mind when facing some particular situation.

We have emphasized the importance of the passage from the statement of a certain optimization problem, often in plain words, to its precise, quantitative formulation that enables us to eventually solve the problem. Scientists and engineers should become experts in this process. A fundamental attitude not to be forgotten when trying to set up a particular problem or reformulate a situation is to insist on reflecting at every stage of this process our original objective, in such a way that the connection between a situation to be solved and its precise formulation is always there. This requires an active attitude with respect to the formulation or reformulation of a particular problem until we have interpreted every aspect of the situation.

To prevent these general comments from being useless, we dare to provide the following recommendations for those facing an optimization problem.

Understanding the optimality criterion. *There should be a very clear statement of the objective and the way in which optimality is to be measured. In particular, the decision about the variables that the cost depends upon and the constraints among them is crucial. One problem can be set up in many different ways, and it is important to discern which might be the most efficient form of the statement. Moreover, it is important to check extreme values of the variables (or other relevant values) and whether the associated cost is co-*

herent with what might be expected. This sort of analysis may often lead to the realization that an error has been made in the statement, and a revision of variables, restrictions, and objective functional should be made.

Understanding the constraints.   *Restrictions linking in different ways the variables of the problem are equally significant. Those can be of a very distinct nature: equalities, inequalities, differential equations, integral restrictions, etc., and may also be hidden in several forms, sometimes in a tacit or implicit manner. What is vital is to analyze the relationship among the variables and the constraints that must be respected. In particular, equalities may be conveniently utilized to decrease the number of variables. The same attitude described above ought to push us to check constraints and their coherence with respect to the situation we want to examine.*

Reflecting on the precise formulation.   *Once the two previous steps have been covered, it is worthwhile to ponder the mathematical formulation of our problem. Do constraints seem coherent? Could the set of feasible vectors or fields be empty? Could some of the restrictions be simplified or eliminated altogether because some constraints are stricter than others? Could the cost be made as small as we like without violating any of the constraints? If so, it is more than likely that we have forgotten some restriction. Could we possibly anticipate whether there is a single optimal solution or whether there could be several?*

Brief analysis of solutions.   *Finally, it is a good thing to get used to examining briefly the optimal solution that has been obtained or approximated. Does it seem like a minimum cost, a maximum efficiency, etc.? Is it plausible that it is indeed an optimal solution? Does it reflect the desired optimality with respect to the terms of the initial problem? Does it satisfy all the requirements?*

As the saying goes, "practice makes perfect," and optimization problems and techniques are no exception. Exercises and situations will help students to go through all the stages described above rapidly and accurately. In the beginning there will be errors, insecurity, inefficiency, shortage of ideas to overcome difficulties, etc., but as students master these aspects, self-confidence will result.

We now proceed to provide the precise formulation of the different problems proposed in the last section. We urge students to work on understanding the connection between the original formulation of a problem and its translation into equations, formulas, inequalities, equalities, etc. This process typically involves setting up a model of the proposed situation. In some simple cases, such a model will be sufficiently clear, and no particular difficulty will be encountered in putting the problem in the appropriate format. In others, however, there may be an initial gap in understanding the mechanisms associated with a specific situation, and additional effort will be necessary to grasp its significance and reach a precise formulation.

The transportation problem. *If x_{ij} is the amount of the product sent from initial location i to destination j, the total cost will be*

$$\sum_{i,j} c_{ij} x_{ij}$$

if c_{ij} is the unit cost of sending the product from i to j. What are the restrictions we must respect? For a fixed service point i, u_i is the quantity to be shipped, so that

$$\sum_j x_{ij} = u_i, \quad i = 1, 2, \ldots, n;$$

likewise, for every fixed destination, the amount v_j should be received, and this enforces

$$\sum_i x_{ij} = v_j, \quad j = 1, 2, \ldots, m.$$

Notice that these two sets of equalities are compatible if

$$\sum_i u_i = \sum_j v_j,$$

which is a restriction that the data of the problem must satisfy for the problem to be well posed. Moreover, if we accept that the feature of being a service point or a destination cannot be reversed, then we must ask for

$$x_{ij} \geq 0, \quad \text{for all } i, j.$$

Altogether, we are seeking to

$$\text{Minimize} \quad \sum_{i,j} c_{ij} x_{ij}$$

under

$$\sum_j x_{ij} = u_i, \quad i = 1, 2, \ldots, n;$$

$$\sum_i x_{ij} = v_j, \quad j = 1, 2, \ldots, m;$$

$$x_{ij} \geq 0, \quad \text{for all } i, j.$$

The diet problem. Let x_i be the amount of food i to be bought. The total cost we would like to minimize is

$$\sum_i c_i x_i$$

if c_i is the unit price of food i. Let a_{ij} be the content of nutrient j per unit of food i, and b_j the daily minimum required of nutrient j. Then we must make sure that in our choice of the diet this minimum is met:

$$\sum_i a_{ji} x_i \geq b_j, \quad \text{for all } j.$$

Finally, we must ask for the nonnegativity of each x_i:

$$x_i \geq 0, \quad \text{for all } i.$$

The problem is

$$\text{Minimize} \quad \sum_i c_i x_i$$

subject to

$$\sum_i a_{ji} x_i \geq b_j, \quad \text{for all } j,$$

$$x_i \geq 0, \quad \text{for all } i.$$

The scaffolding system. If we denote by T_A, T_B, T_C, T_D, T_E, T_F the tensions on each rope when they bear an overall load x_1 and x_2, applied at points x_3 and x_4 units away from the left endpoints of each corresponding beam, the conditions of equilibrium of force and momentum lead to the equations

$$T_E + T_F = x_2, \quad 8T_F = x_4 x_2,$$
$$T_C + T_D = x_1 + T_E + T_F, \quad 10T_D = x_3 x_1 + 2T_E + 10T_F,$$
$$T_A + T_B = T_C + T_D, \quad 12T_B = 2T_C + 12T_D.$$

If we now express the different tensions on each rope in terms of our design variables x_i, we have

$$\frac{x_2 x_4}{8} = T_F \le 100, \quad \frac{8x_2 - x_2 x_4}{8} = T_E \le 100,$$
$$\frac{2x_2 + x_1 x_3 + x_2 x_4}{10} = T_D \le 200, \quad \frac{10x_1 + 8x_2 - x_1 x_3 - x_2 x_4}{10} = T_C \le 200,$$
$$\frac{2x_1 + 4x_2 + x_1 x_3 + x_2 x_4}{12} = T_B \le 300, \quad \frac{10x_1 + 8x_2 - x_1 x_3 - x_2 x_4}{12} = T_A \le 300,$$

and these inequalities should be satisfied. Moreover, we must ask for

$$x_1 \ge 0, \quad x_2 \ge 0, \quad 0 \le x_3 \le 10, \quad 0 \le x_4 \le 8.$$

The problem is then to
$$\text{Maximize} \quad x_1 + x_2$$

subject to

$$x_1 \ge 0, \quad x_2 \ge 0,$$
$$0 \le x_3 \le 10, \quad 0 \le x_4 \le 8,$$
$$x_2 x_4 \le 800, \quad 8x_2 - x_2 x_4 \le 800,$$
$$2x_2 + x_1 x_3 + x_2 x_4 \le 2000, \quad 10x_1 + 8x_2 - x_1 x_3 - x_2 x_4 \le 2000,$$
$$2x_1 + 4x_2 + x_1 x_3 + x_2 x_4 \le 3600, \quad 10x_1 + 8x_2 - x_1 x_3 - x_2 x_4 \le 3600.$$

Power circuit state estimation. In this example, we are told to minimize the mean quadratic error of certain measurements with respect to the predicted values. Specifically, we seek to

$$\text{Minimize} \quad \sum_{i \in \Omega} k_i^v (v_i - \overline{v}_i)^2 + \sum_{i \in \Omega} \sum_{j \in \Omega_i} k_{ij}^p (p_{ij} - \overline{p}_{ij})^2 + \sum_{i \in \Omega} \sum_{j \in \Omega_i} k_{ij}^q (q_{ij} - \overline{q}_{ij})^2$$

where the different data are given in the statement and

$$p_{ij} = \frac{v_i^2}{z_{ij}} \cos \theta_{ij} - \frac{v_i v_j}{z_{ij}} \cos(\theta_{ij} + \delta_i - \delta_j),$$

$$q_{ij} = \frac{v_i^2}{z_{ij}} \sin \theta_{ij} - \frac{v_i v_j}{z_{ij}} \sin(\theta_{ij} + \delta_i - \delta_j).$$

The unknown variables are (v_i, δ_i), and we do not have any explicit restriction on these. Here Ω is the set of nodes, while Ω_i is the set of those connected to node i.

Design of a moving solid. According to our previous explanation and the corresponding diagram, the component along the x-axis of the normal pressure on a point on the surface of the solid is

$$p \sin \theta = 2\rho v^2 \sin^3 \theta.$$

The total pressure in a slice of width dx will be the product of the previous expression times the lateral surface of the slice,

$$dP = 2\rho v^2 \sin^3 \theta \, 2\pi y(x) \sqrt{1 + y'(x)^2} \, dx,$$

if a given profile of the solid is obtained by rotating the graph of the function $y(x)$. If we write $\sin \theta$ in terms of $\tan \theta = y'(x)$, we arrive at

$$dP = 2\rho v^2 2\pi \frac{y'(x)^3}{(1 + y'(x)^2)^{3/2}} y(x) \sqrt{1 + y'(x)^2} \, dx,$$

or simplifying,

$$dP = 4\pi \rho v^2 \frac{y(x) y'(x)^3}{1 + y'(x)^2} \, dx.$$

The objective functional providing the total pressure is

$$P = 4\pi \rho v^2 \int_0^L \frac{y(x) y'(x)^3}{1 + y'(x)^2} \, dx,$$

and we are interested in finding the profile $y(x)$ that minimizes the previous integral among all (continuous) profiles satisfying $y(0) = 0$, $y(L) = R$.

Design of a channel.   Since losses at the wall of a channel are proportional to the inverse of the perimeter, for a given fixed cross section A, the best profile is to be found in the sense that it should have the least perimeter possible. More specifically, we are seeking the profile $y(x)$ such that it minimizes the integral

$$\int_0^R \sqrt{1 + y'(x)^2}\, dx,$$

which provides the length of the graph of $y(x)$, subject to

$$y(0) = 0, \quad y(R) = 0, \quad \int_0^R y(x)\, dx = A.$$

Boat manufacturer.   This problem is self-explanatory, and no further comments are needed.

The harmonic oscillator with friction.   In this example, the best control $u(t)$ is to be found that leads the oscillating surface to rest as soon as possible and at the same time respects the restriction on the size $|u(t)| \leq C$.

A positioning problem.   A mobile object in a plane can be controlled by two parameters at our disposal, r_1 and r_2, expressing the modulus of change of velocity and the rapidity with which the direction of movement can be changed (angular velocity of movement), respectively. The equations of motion are

$$x''(t) = \cos \theta(t) r_1(t), \quad y''(t) = \sin \theta(t) r_1(t), \quad \theta'(t) = r_2(t).$$

Restrictions on the feasible pairs (r_1, r_2) are written by requiring

$$(r_1, r_2) \in [-a, a] \times [-\alpha, \alpha].$$

The objective is to change the position of the object from, say, (x_0, y_0) standing at rest $x'(0) = y'(0) = 0$ at the initial time, to the origin in minimum time

$$x(T) = y(T) = x'(T) = y'(T) = 0,$$

for T as small as possible.

3. THE VARIETY OF OPTIMIZATION PROBLEMS

We have already noted, and it is more than likely that readers have also appreciated it, the tremendous differences among optimization problems. These differences have motivated the structure of this text.

Perhaps the most significant difference lies in the fact that in some problems, vectors describe solutions and optimal solutions, whereas in other cases functions are needed to formulate and solve the problem. This important, profound qualitative distinction results in a difference between optimization techniques for these two categories of problems. The situation is similar to the case of equations or systems of equations in which we are interested in a vector solution, a bunch of numbers, and differential equations where the unknown is a function. In the first case, we talk about mathematical programming; in the second, about variational problems. In a second approximation, mathematical programming can be divided into linear programming (Chapter 2), dealing with the simpler world of linear problems, and nonlinear programming (Chapter 3), for the complex nonlinear optimization techniques. The transportation and diet problems correspond to linear programming, while the scaffolding system and the power circuit state estimation are examples of nonlinear optimization problems.

The type of situations where we intend to find optimal functions for specific situations can be classified into variational problems (Chapter 5) with a brief incursion into dynamic programming, and optimal control problems (Chapter 6). The design of a moving solid or a channel and the boat manufacturer problem correspond to variational problems and dynamic programming. The harmonic oscillator and the positioning problem are typical examples of optimal control problems.

Chapter 4 is like a point of intersection between the world of vectors and that of functions. We will understand this assertion later. Our aim in this chapter is to describe the most basic and relevant numerical algorithms for computing and/or approximating optimal solutions to problems. Since in most of the real situations one may encounter, exact optimal solutions are not to be expected, these computational techniques are crucial. We will restrict attention to the most basic, well-known such techniques. Our objective is to let readers have some idea about the nature of approximation techniques for optimization problems. We have not included explicit implementation of algorithms for two main reasons. There are a number of existing and tested commercial optimization software packages (see Chapter 4 for some specific references) that are quite helpful, since they free us from having to be concerned about technical issues related to approximation, and instead focus on the modeling task. On the other hand, the fine tuning of algorithms, especially when nonlinear restrictions must be taken into account, requires considerable experience and expertise as soon as the number of independent variables grows above a few. The nonexpert would probably do a poor job compared to that carried out in those software packages. This does not mean that it is useless to have some experience trying to write personal programs for some simple situations. We have written down some simple versions of algorithms in pseudocode format.

Finally, it is important to stress that each of these chapters is but a timid initiation into the corresponding ideas. The wealth of situations, the peculiarities of realistic problems, the need for better computational methods and algorithms, and the need for a deeper understanding of the structure of problems can be such that a whole book would be needed to more fully cover each of these small chapters. Our intention is to furnish a first overall view of optimization, emphasizing the basic ideas and techniques in each category of optimization problem.

4. EXERCISES

1. An investor is seeking to invest a certain capital K in a diversified manner so as to maximize expected profits at the end of a certain period of time. If r_i is the expected average interest rate for investment i, and to avoid excesive risk he (she) does not want to put on any one investment more than a fixed percentage r of the capital, formulate the problem leading to the best solution. Can you figure out other types of reasonable restrictions to enforce in such a situation?

2. In the context of the scaffolding system described earlier in the chapter, assume that the points where loads x_1 and x_2 are applied are exactly the midpoints of beams CD and EF, respectively. Formulate the problem. What is the main difference between this situation and the one described in the text?

3. A company that manufactures tiling elements for roofs must provide 7800 m^2 of these elements for several houses. Two different elements can be used: Model A10 requires 9.5 elements per square meter, and model A13 needs

12.5 elements per square meter. Both models can be used in the same roof. The respective prices are 0.70 and 0.80 euros per element. The company has 1600 labor hours to finish the roofs. In one hour, 5 m^2 of model A10 and 4 m^2 of model A13 can be installed. Due to baking restrictions, the maximum amount of model A13 that can be sent is 2500 m^2. Formulate the problem of maximizing benefits subject to all of the restrictions indicated.

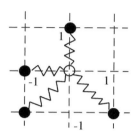

Figure 1.5. A system of springs.

4. In the system of springs of Figure 1.5, each node is free to rotate about itself. If each spring has a constant k_i characterizing elongation (according to Hooke's law) and the equilibrium position of the free central node is determined through the system

$$\sum_i k_i(x - x_i) = 0,$$

where x_i is the position of fixed joints, describe how to determine the optimal spring constants k_i that minimize the work done by a constant force F on the free node, assuming that

$$\sum_i k_i = k,$$

a fixed positive constant.

5. A company is to build several (m) service points to serve a certain number (n) of known clients. A decision is to be made about the optimal location of those service points. Assuming that the criterion chosen is global minimal

distance from service points to clients, state the problem as a nonlinear programming problem. Describe other ways of making that decision.

6. A quadrature formula is a way to efficiently approximate definite integrals through sums of the type

$$\int_a^b f(x)\,dx \approx \sum_{j=1}^n \alpha_j f(x_j),$$

where weights α_j and points x_j determine the particular quadrature rule. We would like to determine the vector of n weights (α_j) and n points (x_j) in the interval $[-1,1]$ so that the corresponding quadrature is exact for polynomials of degree as high as possible. The procedure is to minimize the quadratic error of the quadrature formula for polynomials of degree m. State the problem as a nonlinear programming problem.

7. The Cobb–Douglas utility function is of the form

$$u(x,y) = x^\alpha y^{1-\alpha}, \quad 0 < \alpha < 1, \quad x \ge 0, \quad y \ge 0.$$

Assume an economy of two consumers, 1 and 2, and two commodities X and Y. Both consumers have the same utility function of the type above with the same exponent α, and resources

$$(\overline{x}_i, \overline{y}_i), \quad i = 1, 2,$$

for each commodity. If prices $p = (p_X, p_Y)$ prevail in the market for both commodities, formulate the problem of maximizing satisfaction for each consumer as measured by their utility functions.

8. A ladder must lean against a wall where a box of dimensions $a \times b$ is placed against the same wall as in Figure 1.6. Formulate the problem of finding the shortest such ladder.

9. John is supposed to cut n_i bars of length a_i, $i = 1, 2, \ldots, d$, from bars of fixed given length L, $a_i \le L$ for all i. What is the minimum number of such bars he needs? Find a precise formulation of this optimization problem.

10. An airplane is flying with speed v with respect to the ground in a bounded irrotational wind field given by $\nabla\varphi(x, y, z)$ and such that $v > |\nabla\varphi|$. Starting and ending at the same point, what are the longest and shortest paths it can fly in a given time interval $[0, T]$? Write down the problem assuming

that v is constant and the direction of velocity is at our disposal. (Hint: Write a parametrization of a curve

$$\sigma(t) = (x(t), y(t), z(t)).$$

What do we know about x', y', z' in terms of v, the direction of velocity and $\nabla\varphi$? Keep in mind that the length of such a curve is given by

$$\int_0^T |\sigma'| \, dt.)$$

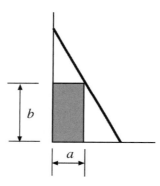

Figure 1.6. A ladder against a wall.

11. A rope is hanging vertically in equilibrium from its upper fixed endpoint (Figure 1.7). It is stretched by the action of its own weight and a constant mass W at its lower end. The problem consists in determining the optimum distribution of the cross-sectional area $a(x)$, $0 \le x \le L$, so as to minimize the total elongation. The unstretched length L, the total volume V, the density ρ, and Young modulus E are constant and known.
   1. What is the integral restriction related to the volume V that the function $a(x)$ must satisfy to be admissible?
   2. Let $y(x)$ be the distance, measured form the upper fixed endpoint and corresponding to the design $a(x)$, that the section at distance x in the unstretched configuration moves to when the rope is pulled by the weight

W. Assume that Hooke's law applies: The strain $y'(x)$ at each point is proportional (with proportionality constant $1/E$) to the stress there, where the stress at x is the total downward force divided by the cross-sectional area $a(x)$. Write down this law in the form of an equation.

3. How is the objective expressed in terms of y? Is there a further restriction to be imposed on y?

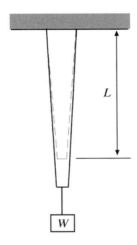

Figure 1.7. A rope with varying cross section.

12. The problem of the slowest descent to the moon can be formulated in the following terms. If $v(t)$ and $m(t)$ are the velocity and combined mass of the spacecraft and fuel at time t, σ is the (constant) relative ejection velocity of fuel, and g is gravity, then the state law is written

$$(m + dm)(v + dv) - dm\,(v + \sigma) - mv = mg\,dt,$$

or equivalently,

$$\frac{dv}{dt} = g + \frac{\sigma}{m}\frac{dm}{dt}.$$

If the rate of ejection per unit time $-dm/dt$ can be controlled within an interval $[0, \alpha]$, formulate the problem of soft landing in minimum time in precise terms.

13. A jet plane is to reach a certain point in space in minimum time from take-off. Assuming that the total energy (kinetic plus potential plus (minus) fuel) is constant, the jet burns fuel at its maximum constant rate, and it has zero velocity at takeoff, formulate the corresponding optimization problem. (Hint: The equation of total energy leads us to postulate

$$v^2 + 2gy = at,$$

where $v = (x', y')$ is the velocity, g is the acceleration due to gravity, and a is the constant maximum rate at which the jet burns fuel.)

14. In connection with the construction of an optimal refracting medium, the following problem arises:

$$\text{Maximize} \quad y(1)$$

subject to

$$y''(x) - F(x)y(x) = 0, \quad y(0) = 1, \quad y'(0) = 0,$$

$$F \geq 0, \quad \int_0^1 F(x)\, dx = M.$$

Reformulate this problem as an optimal control problem with an integral objective functional.

15. An aggregate model of economic growth can be described by the following equations

$$Y(t) = F(L(t), K(t)),$$

$$K'(t) + \mu K(t) = Y(t) - X(t), \quad \frac{L'(t)}{L(t)} = n,$$

where Y is the single output of the economy, using two inputs, labor (L) and capital (K), X denotes the amount of consumption, μ is the rate of depreciation, the variable t indicates time, and n is the constant rate at which labor grows. The objective of this economy is to maximize the welfare integral

$$\int_0^\infty u(X(t)/L(t))e^{-\rho t}\, dt,$$

where ρ is the time discount factor. Try to simplify the formulation of this problem as much as possible.

16. Sometimes, optimization problems may not adapt themselves to either of the formats described in this chapter, mainly because the optimality criterion is more involved than the ones envisioned in this text. For example, a hydraulic cushion unit (Figure 1.8), such as those used in the railroad industry, develops a cushioning force given by

$$F = c\frac{v^2}{a^2}, \quad 0 \le x \le x_m,$$

where c is a constant, $v = v(x)$ is the velocity of the cushion, $a(x)$ is an orifice area that is allowed to vary with displacement x, and x_m is the maximum displacement permitted under appropriate geometric constraints. The design of such units seeks to choose $a(x)$ so as to minimize the maximum force for a given impact mass m with impacting velocity v_0. Show that the optimum is obtained when $a(x)^2$ varies linearly with x. (Hint: The work energy formula is

$$\frac{1}{2}mv^2 = \frac{1}{2}mv_0^2 - \int_0^x F(s)\,ds.$$

What information does this equation provide at the end of impact when $v = 0$?)

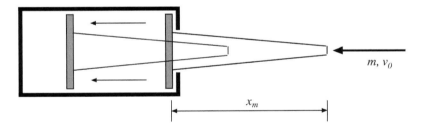

Figure 1.8. A cushion unit.

Chapter 2

Linear Programming

1. INTRODUCTION

The main feature of a linear programming problem (LPP) is that all functions involved, the objective function and those expressing the constraints, must be linear. The appearance of a single nonlinear function, either on the objective or in the constraints, suffices to reject the problem as an LPP.

Definition 2.1 *(General form of an LPP) An LPP is an optimization problem of the general form*

$$\text{Minimize} \quad cx = \sum_i c_i x_i$$

subject to

$$\sum_i a_{ji} x_i \le b_j, \quad j = 1, \ldots, p,$$

$$\sum_i a_{ji} x_i \ge b_j, \quad j = p+1, \ldots, q,$$

$$\sum_i a_{ji} x_i = b_j, \quad j = q+1, \ldots, m,$$

where c_i, b_j, a_{ji} are data of the problem. Depending on the particular values of p and q we may have inequality constraints of one type and/or the other, and equality restrictions as well.

We can gain some insight into the structure and features of an LPP by looking at one simple example.

Example 2.2   *Consider the LPP*

$$\text{Maximize} \quad x_1 - x_2$$

subject to

$$x_1 + x_2 \le 1, \quad -x_1 + 2x_2 \le 2,$$
$$x_1 \ge -1, \quad -x_1 + 3x_2 \ge -3.$$

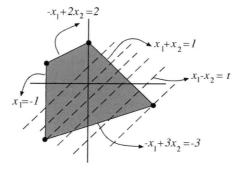

Figure 2.1. The feasible region and level curves in an LPP.

It is interesting to realize the shape of the set of vectors in the plane satisfying all the requirements that the constraints express: Each inequality represents a "half-space" at one side of the line corresponding to changing the inequality to an equality. Thus the intersection of all four half-paces will be the "feasible region" for our problem. Notice that this set has the form of a polygon or polyhedron. See Figure 2.1.

On the other hand, the cost, being linear, has level curves that are again straight lines of equation $x_1 - x_2 = t$, a constant. When t moves, we obtain parallel lines. The question is then how big t can become so that the line of equation $x_1 - x_2 = t$ meets the above polygon somewhere. Graphically, it is not hard to realize that the optimal vector corresponds to the vertex $(-1/2, 3/2)$, and the value of the maximum is 2.

Note that regardless of what the cost is, as long as it is linear, the optimal value will always correspond to one of the four vertices of the feasible set. These vertices play a crucial role in the understanding of LPP, as we will see.

An LPP can adopt several equivalent forms. The initial form usually depends on the particular formulation of the problem, or the most convenient way in which the constraints can be represented. The fact that all possible formulations correspond to the same underlying optimization problem enables us to fix one reference format, and refer to this form of any particular problem for its analysis.

Definition 2.3   (Standard form of an LPP) An LPP in standard form is

$$\text{Minimize} \quad cx \quad \text{under} \quad Ax = b, \quad x \geq 0. \tag{P}$$

Thus, the ingredients of every LPP are:
1. an $m \times n$ matrix A, with $n > m$ and typically n much greater than m;
2. a vector $b \in \mathbf{R}^m$;
3. a vector $c \in \mathbf{R}^n$.

Notice that cx is the inner product of the two vectors c and x, while Ax is the product of the matrix A and the vector x. We will not make the distinction between these possibilities, since it will be clear from the context. It is therefore a matter of finding the minimum value the inner product cx can take on as x runs through all feasible vectors $x \in \mathbf{R}^n$ with nonnegative components ($x \geq 0$) satisfying the additional, and important restriction $Ax = b$. We are also interested in one vector x (or all vectors x) where this minimum value is achieved.

We have argued that any LPP can in principle be transformed into the standard form. It is therefore desirable that readers understand how this transformation can be accomplished. We will proceed in three steps.

1. Variables not restricted in sign. For the variables not restricted in sign, we use the decomposition into positive and negative parts according to the identities

$$x = x^+ - x^-, \quad |x| = x^+ + x^-,$$

where

$$x^+ = \max\{0, x\} \geq 0, \quad x^- = \max\{0, -x\} \geq 0.$$

What we mean with this decomposition is that a variable x_i not restricted in sign can be written as the difference of two new variables that are nonnegative:

$$x_i = x_i^{(1)} - x_i^{(2)}, \quad x_i^{(1)}, x_i^{(2)} \geq 0.$$

2. Transforming inequalities into equalities. Quite often, restrictions are formulated in terms of inequalities. In fact, an LPP will come many times in the form

$$\text{Minimize} \quad cx \quad \text{under} \quad Ax \leq b, \quad A'x = b', \quad x \geq 0.$$

Notice that by using multiplication by minus signs we can change the direction of an inequality. In this situation, the use of "slack variables" permits the passage from inequalities to equalities in the following way. Introduce new variables by putting

$$y = b - Ax \geq 0.$$

If we now set

$$X = (x \quad y), \quad \tilde{A} = (A \quad 1),$$

where 1 is the identity matrix of the appropriate size, the inequality restrictions are written now as

$$\tilde{A}X = b,$$

so all constraints are now in the form of equalities, but we have a greater number of variables (one more for each inequality).

3. Transforming a max into a min. If the LPP asks for a maximum instead of for a minimum, we can keep in mind that

$$\max(\text{expression}) = -\min(-\text{expression});$$

or more explicitly,

$$\max\{cx : Ax = b, x \geq 0\} = -\min\{(-c)x : Ax = b, x \geq 0\}.$$

An example will clarify any doubt about these transformations.

Example 2.4 *Consider the LPP*

$$\text{Maximize} \quad 3x_1 - x_3$$

subject to

$$x_1 + x_2 + x_3 = 1,$$
$$x_1 - x_2 - x_3 \leq 1,$$
$$x_1 + x_3 \geq -1,$$
$$x_1 \geq 0, \quad x_2 \geq 0.$$

1. Since there are variables not restricted in sign, we must set

$$x_3 = y_1 - y_2, \quad y_1 \geq 0, y_2 \geq 0,$$

so that the problem will change to

$$\text{Maximize} \quad 3x_1 - y_1 + y_2$$

subject to

$$x_1 + x_2 + y_1 - y_2 = 1,$$
$$x_1 - x_2 - y_1 + y_2 \leq 1,$$
$$x_1 + y_1 - y_2 \geq -1,$$
$$x_1 \geq 0, \quad x_2 \geq 0,$$
$$y_1 \geq 0, \quad y_2 \geq 0.$$

2. We use slack variables so that inequality restrictions may be transformed into equalities: $z_1 \geq 0$ and $z_2 \geq 0$ are used to transform

$$x_1 - x_2 - y_1 + y_2 \leq 1, \quad x_1 + y_1 - y_2 \geq -1,$$

respectively, into

$$x_1 - x_2 - y_1 + y_2 + z_1 = 1, \quad z_1 \geq 0,$$

and

$$x_1 + y_1 - y_2 - z_2 = -1, \quad z_2 \geq 0.$$

The problem will now have the form

$$\text{Maximize} \quad 3x_1 - y_1 + y_2$$

subject to

$$x_1 + x_2 + y_1 - y_2 = 1,$$
$$x_1 - x_2 - y_1 + y_2 + z_1 = 1,$$
$$x_1 + y_1 - y_2 - z_2 = -1,$$
$$x_1 \geq 0, \quad x_2 \geq 0,$$
$$y_1 \geq 0, \quad y_2 \geq 0,$$
$$z_1 \geq 0, \quad z_2 \geq 0.$$

3. Finally, we easily change the maximum to a minimum:

$$\text{Minimize} \quad -3x_1 + y_1 - y_2$$

subject to

$$x_1 + x_2 + y_1 - y_2 = 1,$$
$$x_1 - x_2 - y_1 + y_2 + z_1 = 1,$$
$$x_1 + y_1 - y_2 - z_2 = -1,$$
$$x_1 \geq 0, \quad x_2 \geq 0,$$
$$y_1 \geq 0, \quad y_2 \geq 0,$$
$$z_1 \geq 0, \quad z_2 \geq 0,$$

bearing in mind that once the value of this minimum is found, the corresponding maximum will have its sign changed.

If we uniformize the notation by writing

$$(X_1, X_2, X_3, X_4, X_5, X_6) = (x_1, x_2, y_1, y_2, z_1, z_2),$$

the problem will obtain its standard form

$$\text{Minimize} \quad X_3 - X_4 - 3X_1$$

subject to

$$X_1 + X_2 + X_3 - X_4 = 1,$$
$$X_1 - X_2 - X_3 + X_4 + X_5 = 1,$$
$$X_1 + X_3 - X_4 - X_6 = -1,$$
$$X \geq 0.$$

Once this problem has been solved and we have an optimal solution X and the value of the minimum m, the answer to the original LPP would be as follows: The maximum is $-m$, and it is achieved at the point $(X_1, X_2, X_3 - X_4)$. Or if you like, the value of the maximum will be the value of the original linear cost function at the optimal solution $(X_1, X_2, X_3 - X_4)$. Notice how the slack variables do not enter into the final answer, since they are auxiliary variables.

Concerning the optimal solution of an LPP, all situations can actually happen:

1. the set of admissible vectors is empty;
2. it can have no solution at all, because the cost cx can decrease indefinitely toward $-\infty$ for feasible vectors x;
3. it can admit a single optimal solution, and this is the most desirable situation;
4. it can also have several, in fact infinitely many, optimal solutions; indeed, it is very easy to check that if x_1 and x_2 are optimal, then any convex combination

$$tx_1 + (1-t)x_2, \quad t \in [0,1],$$

is again an optimal solution.

In the next section, we will see how to solve an LPP in its standard form by the simplex method. Though interior-point methods are becoming more and more important in mathematical programming, in both versions, linear and nonlinear, we tend to believe that they are the subject of a second course on mathematical programming. The fact is that the simplex method helps greatly in understanding the special structure of linear programming as well as duality.

2. THE SIMPLEX METHOD

We look more closely at an LPP in its standard form, and describe the simplex method, which is one of the most successful approaches for finding the optimal solution for such problems. Let us concentrate, then, on the problem of finding a vector x solving

$$\text{Minimize} \quad cx$$

subject to

$$Ax = b, \quad x \geq 0.$$

There is no restriction in assuming that the linear system $Ax = b$ is solvable, for otherwise, there would be no feasible vectors. Moreover, if A is not a full-rank matrix, we can select a submatrix A' by eliminating several rows of A, and the corresponding components of b, so that the new matrix A' has full rank. In this case we obtain the new, equivalent, LPP

$$\text{Minimize} \quad cx$$

subject to

$$A'x = b', \quad x \geq 0,$$

where b' is the subvector of b obtained by eliminating the components corresponding to the rows of A we have previously discarded. This new LPP is equivalent to the initial one in the sense that they both have the same optimal solutions, but the matrix A' for the reduced problem is a full-rank matrix. We shall therefore assume, without loss of generality, that the rank of the $m \times n$ matrix A is m (remember $m \leq n$) and that the linear system $Ax = b$ is solvable.

There are special feasible vectors that play a central role in the simplex method. These are the solutions of the linear system $Ax = b$ with nonnegative and at least $n - m$ null components. In fact, all of these extremal points or basic solutions, as they are typically called, can, in principle, be obtained by solving all square $m \times m$ linear systems $Ax = b$ where $n - m$ components of x are set to zero, and discarding those solutions with at least one negative component. The very special linear structure of an LPP enables us to concentrate on these basic solutions when looking for optimal solutions.

Lemma 2.5   *If the LPP*

$$\text{Minimize} \quad cx$$

subject to

$$Ax = b, \quad x \geq 0,$$

admits an optimal solution, then there is also an optimal solution that is a basic solution.

This is quite evident if we realize that the feasible set of an LPP is some kind of "polyhedron," and therefore minimum (or maximum) values of linear functions must be taken on at a vertex. See Figure 2.2, and remember the comments on Example 2.2.

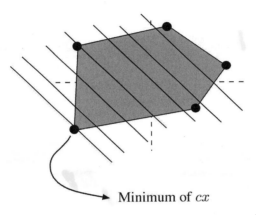

Minimum of cx

Figure 2.2. Optimal basic solution.

For the proof of the lemma, assume that x is an optimal solution with at least $m + 1$ strictly positive components, and let d be a nonvanishing vector in the kernel of A with the property that $x_i = 0$ implies $d_i = 0$. If x has at least $m + 1$ strictly positive components, such a vector d can always be found (why?).

We claim that necessarily $cd = 0$. For otherwise, if t is small enough so that $x + td$ is feasible (i.e., $x + td \geq 0$) and $tcd < 0$, then the cost of the vector $x + td$ is stricly smaller than that of x itself, which is impossible if x is optimal. Therefore, $cd = 0$, and the vectors $x + td$ are also optimal as long as they are feasible. All that remains to be done is to move t away from zero (either positive or negative) until some of the components of $x + td$ hit zero for the first time. For such value of t we would have an optimal solution with at least one more vanishing component than x. This process can be repeated as long as the vector d is not the zero vector, i.e., until the optimal solution has at least $n - m$ vanishing components.

As an immediate consequence of Lemma 2.5, we can find optimal solutions for an LPP by looking at all solutions of the system $Ax = b$ with at least $n - m$ zeros, discarding those with some negative component, and, by computing the cost of the remaining ones, decide on the optimal vector. This process would indeed lead us to one optimal solution, but the simplex method aims to organize these computations in a judicious way so that we can reach the optimal solution as soon as possible without having to go through an exhaustive analysis of all extremal points. In some cases, though, the simplex method actually goes through all basic solutions before finding an extremal solution. This situation is, however, rare.

The SM starts at one particular extremal feasible vector x, which, after an appropriate permutation of indices, can be written as

$$x = (\,x_B \quad 0\,), \quad x_B \in \mathbf{R}^m, \quad 0 \in \mathbf{R}^{n-m}, \quad x_B \geq 0.$$

The basic iterative step consists in setting one of the components of x_B to zero (the so-called "leaving variable"), and letting a vanishing component of $0 \in \mathbf{R}^{n-m}$ (the "entering variable") become positive. In this way, we have moved from an extremal point to an adjacent one. The key issue is to understand how to make these choices (leaving and entering variables) in such a way that we lower the cost as much as possible. Furthermore, we need a criterion to decide when the cost cannot be decreased any more, so that we have actually found an optimal solution and no more iterative steps are needed. We discuss this procedure more precisely in what follows.

Let

$$x = (\,x_B \quad 0\,), \quad x_B \in \mathbf{R}^m, \quad 0 \in \mathbf{R}^{n-m}, \quad x_B \geq 0,$$

be a feasible extremal vector. In the same way, the matrix A, after the same permutation of columns, can be decomposed as $B = $ basic variables

$$A = (\,B \quad N\,). \quad N = \text{non basic variable}$$

The equation $Ax = b$ is equivalent to

$$(\,B \quad N\,) \begin{pmatrix} x_B \\ 0 \end{pmatrix} = b, \quad x_B = B^{-1}b.$$

The cost of such a vector x is

$$cx = \begin{pmatrix} c_B & c_N \end{pmatrix} \begin{pmatrix} x_B \\ 0 \end{pmatrix} = c_B x_B = c_B B^{-1} b.$$

The basic step of the simplex method consists in moving to another feasible (adjacent) extremal point so that the cost has been lowered in such a movement. The change from $x = \begin{pmatrix} x_B & 0 \end{pmatrix}$ to $\overline{x} = \begin{pmatrix} \overline{x}_B & x_N \end{pmatrix}$, where x_N is at our disposal, will take place if we can ensure three requirements:

The first one forces us to take

$$\overline{x}_B = x_B - B^{-1} N x_N.$$

Indeed, notice that

$$\begin{pmatrix} B & N \end{pmatrix} \begin{pmatrix} \overline{x}_B \\ x_N \end{pmatrix} = b$$

implies

$$\overline{x}_B = B^{-1}(b - N x_N) = x_B - B^{-1} N x_N.$$

Consequently, the new cost will be

$$\begin{pmatrix} c_B & c_N \end{pmatrix} \begin{pmatrix} x_B - B^{-1} N x_N \\ x_N \end{pmatrix} = c_B(x_B - B^{-1} N x_N) + c_N x_N$$

$$= (c_N - c_B B^{-1} N) x_N + c_B x_B.$$

We see that the sign of

$$(c_N - c_B B^{-1} N) x_N$$

will dictate whether we have been able to decrease the cost by moving to the new vector

$$\begin{pmatrix} x_B - B^{-1} N x_N & x_N \end{pmatrix}.$$

The so-called vector of reduced costs

$$r = c_N - c_B B^{-1} N \qquad (\text{row } 0)$$

will play an important role in deciding whether we can move to a new basic solution and lower the cost. Since $x_N \geq 0$ (by requirement 3 above), two situations may occur:

1.Stopping criterion. If all components of r turn out to be nonnegative, there is no way to lower the cost, and the present extremal point is indeed optimal. We have found a solution for our problem.

2. Iterative step. If r has some negative components, we can, in principle, lower the cost by letting those components of x_N become positive. However, we must exercise caution in this change in order to ensure that the vector

$$x_B - B^{-1}Nx_N \tag{2-1}$$

is feasible, i.e., will always have nonnegative coordinates. If this is not the case, even though the cost will have a smaller value in the vector

$$(\, x_B - B^{-1}Nx_N \quad x_N \,),$$

it will not be feasible and therefore wil not be admissible as an optimal solution of the LPP. We must ensure the nonnegativity of the extremal vectors.

Instead of looking for more general choices of x_N, the simplex method focuses on taking $x_N = tv$, where $t \geq 0$ and v is a basis vector having vanishing coordinates in all but one place, where it has 1. This means that we will change one component at a time. The chosen component is precisely the "entering variable." How is this variable to be selected? According to our previous discussion, we are trying to ensure that the product

$$rx_N = trv$$

will be as negative as it possibly can. Since v is a basis vector, rv is a component of r, and therefore v must be chosen as the basis vector corresponding to the most negative variable of r. Once v has been selected, we have to examine

$$x_B - B^{-1}Nx_N = B^{-1}b - tB^{-1}Nv \tag{2-2}$$

in order to determine the leaving variable. The idea is the following. When $t = 0$, we are sitting on our basic solution x_B. What can happen if t starts to move away from zero to the positive part? At this point three situations may occur. We discuss them succesively.

1. Infeasible solution. As soon as t becomes positive, the vector in (2–2) is not feasible any longer because one of its components is less than zero. In this case, we cannot use the chosen variable to reduce the cost, and we must turn to the next negative variable in r; or alternatively, we can simply take this variable as the leaving variable in spite of the fact that the cost will not decrease. This second choice is usually preferred due to coherence of the whole process.

2. Leaving variable. There is a positive threshold value of t at which one of the coordinates of (2–2) vanishes for the first time. We choose precisely this one as the leaving variable, and compute a new extremal point with smaller cost than the previous one.

3. No solution. No matter how big t becomes, we can lower and lower the cost, and none of the components in (2–2) will ever reach zero. The problem does not admit an optimal solution because we can reduce the cost indefinitely.

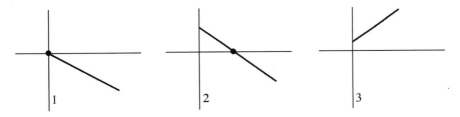

Figure 2.3. Three possibilities in choosing the leaving variable.

The issue is how we can decide in each particular situation whether we are in case 1, 2, or 3, above, and how to proceed accordingly. Notice that each expression in (2–2) represents a straight line as a function of t. The three possibilities are drawn in Figure 2.3.

Assume that we have chosen an entering variable identified with a basis vector v. We proceed as follows:

1. If there is one vanishing component of $x_B = B^{-1}b$ corresponding to a positive component of $B^{-1}Nv$ (diagram 1 in Figure 2.3), then as soon as t becomes positive this coordinate will be smaller than zero in (2–2), and the vector will not be feasible. We might resort to a different entering variable (a different basis vector v), which corresponds to another negative component of r, if available. If r does not have more negative components, we already

have the optimal solution, and the simplex method stops. Alternatively, and this choice is typically preferred for coherence, we can consider the vanishing ratio as one candidate for the process in 2 below.

2. Examine the ratios of the vectors $B^{-1}b$ over $B^{-1}Nv$ componentwise, and choose as leaving variable the one corresponding to the least of those ratios among the strictly positive ones, including, as remarked earlier, the vanishing ratios with positive denominators. These would certainly be chosen, if present, since they are smaller than the strictly positive ones. Disregard the quotients over zero including $0/0$. Start the whole process with the new extremal vector. Notice that these ratios correspond to the values of t when t intersects the horizontal axis in diagram 2 of Figure 2.3.

3. If there are no positive ratios, the LPP does not admit an optimal solution, since the cost can be indefinitely lowered by increasing the entering variable. This situation occurs when all diagrams are of the type 3 in Figure 2.3.

Since the set of feasible extremal points is finite, after a finite number of steps, the simplex method leads us to an optimal solution or to the conclusion that there is no optimal vector. In some very peculiar instances, the simplex method can enter a cyclic, infinite process. Such cases are so rare that we will pay no attention to them. One easy example is proposed as an exercise at the end of the chapter.

In practice, the computations can be organized in the following algorithmic fashion.

1. Initialization.   *Find a square $m \times m$ submatrix B such that the solution of the linear system $Bx_B = b$ is such that $x_B \geq 0$.*

2. Stopping criterion.   *Write*

$$c = (c_B \quad c_N), \quad A = (B \quad N).$$

Solve

$$z^T B = c_B$$

and look at the vector

$$r = c_N - z^T N. \quad (\text{row } 0)$$

If $r \geq 0$, stop: We already have an optimal solution. If not, choose the entering variable corresponding to the most negative component of r.

3. Main iterative step. *Solve*

$$Bw = y,$$

where y is the entering column of N corresponding to the entering variable, and look at the ratios x_B/w componentwise. Among these ratios select those with positive denominators. Choose as leaving variable the one corresponding to the smallest ratio among the selected ones. Go to step 2. If there is no variable to select from, the problem does not admit an optimal solution.

We have tried to reflect the main iterative step of the simplex method in Figure 2.4.

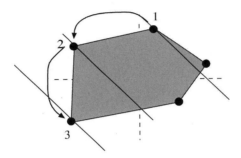

Figure 2.4. Several iterative steps in the simplex method.

In order to ensure that our readers understand the strategy in the simplex method, how the entering and leaving variables are chosen and the stopping criterion, we are going to look briefly at several simple examples.

Example 2.6 *(Unique solution)*

$$\text{Minimize} \quad 3x_1 + x_2 + 9x_3 + x_4$$

subject to

$$x_1 + 2x_3 + x_4 = 4$$
$$x_2 + x_3 - x_4 = 2,$$
$$x_i \geq 0.$$

In this particular instance,

$$A = \begin{pmatrix} 1 & 0 & 2 & 1 \\ 0 & 1 & 1 & -1 \end{pmatrix}, \quad b = \begin{pmatrix} 4 \\ 2 \end{pmatrix}, \quad c = (3 \quad 1 \quad 9 \quad 1).$$

1. Initialization. Choose

$$B = \begin{pmatrix} 1 & 0 \\ 0 & 1 \end{pmatrix}, \quad N = \begin{pmatrix} 2 & 1 \\ 1 & -1 \end{pmatrix}, \quad c_B = (3 \quad 1), \quad c_N = (9 \quad 1).$$

2. Checking the stopping criterion. It is trivial to find

$$x_B = b = \begin{pmatrix} 4 \\ 2 \end{pmatrix}$$

such that the initial vertex is $(4, 2, 0, 0)$ with cost 14. On the other hand,

$$z = c_B = (3 \quad 1), \quad r = (9 \quad 1) - (3 \quad 1)\begin{pmatrix} 2 & 1 \\ 1 & -1 \end{pmatrix} = (2 \quad -1).$$

Since not all components of r are nonnegative, we must go through the iterative process in the simplex method.

3. Iterative step. Choose x_4 as the entering variable, since this is the one associated with the negative component of r. Moreover,

$$\frac{4}{1} = 4 \quad \frac{2}{-1} = -2 \quad w = \begin{pmatrix} 1 \\ -1 \end{pmatrix}, \quad \frac{x_B}{w} = \{4, -2\},$$

so that x_1 is the leaving variable, being the one corresponding to the least ratio among the ones we would select (ratios with positive denominators).

4. Checking the stopping criterion. These computations lead us to the new choice

$$B = \begin{pmatrix} 0 & 1 \\ 1 & -1 \end{pmatrix}, \quad N = \begin{pmatrix} 1 & 2 \\ 0 & 1 \end{pmatrix}, \quad c_B = (1 \quad 1), \quad c_N = (3 \quad 9).$$

It is easy to find

$$x_B = \begin{pmatrix} 6 \\ 4 \end{pmatrix}$$

and the new extremal vector $(0, 6, 0, 4)$ with associated cost 10. The new vectors z and r are

$$z = (2 \quad 1), \quad r = (3 \quad 9) - (2 \quad 1)\begin{pmatrix} 1 & 2 \\ 0 & 1 \end{pmatrix} = (1 \quad 4).$$

Since all components of r are nonnegative, we have ended our search: The minimum cost is 10, and it is taken on at the vector $(0, 6, 0, 4)$.

Example 2.7   (Degenerate example)

$$\text{Minimize} \quad 3x_1 + x_2 + 9x_3 + x_4$$

subject to

$$x_1 + 2x_3 + x_4 = 0,$$
$$x_2 + x_3 - x_4 = 2,$$
$$x_i \geq 0.$$

In this particular case,

$$A = \begin{pmatrix} 1 & 0 & 2 & 1 \\ 0 & 1 & 1 & -1 \end{pmatrix}, \quad b = \begin{pmatrix} 0 \\ 2 \end{pmatrix}, \quad c = (3 \quad 1 \quad 9 \quad 1).$$

1. Initialization. Choose

$$B = \begin{pmatrix} 1 & 0 \\ 0 & 1 \end{pmatrix}, \quad N = \begin{pmatrix} 2 & 1 \\ 1 & -1 \end{pmatrix}, \quad c_B = (3 \quad 1), \quad c_N = (9 \quad 1).$$

2. Checking the stopping criterion. It is trivial to find

$$x_B = b = \begin{pmatrix} 0 \\ 2 \end{pmatrix}$$

such that the initial vertex is $(0, 2, 0, 0)$ with cost 2. On the other hand,

$$z = c_B = (3 \quad 1), \quad r = (9 \quad 1) - (3 \quad 1)\begin{pmatrix} 2 & 1 \\ 1 & -1 \end{pmatrix} = (2 \quad -1).$$

Since not all components of r are nonnegative, we must go through the iterative process in the simplex method.

3. *Iterative step. Choose x_4 as the entering variable, since it is the one associated with the negative component of r. Moreover,*

$$w = \begin{pmatrix} 1 \\ -1 \end{pmatrix}, \quad \frac{x_B}{w} = \{0, -2\},$$

so that x_1 is the leaving variable, being the one corresponding to the least ratio among the ones we would select (ratios with positive denominators). We can predict, however, that because our only choice is a vanishing ratio, we will not be able to lower the cost in spite of going through an iterative step of the simplex method. In other words, the vertex $(0, 2, 0, 0)$ is already an optimal solution. Since for this optimal solution the stopping criterion does not hold for our original choice of B (r has negative coordinates), we must, for the sake of coherence of the scheme, go through an iterative step of the simplex method.

4. *Checking the stopping criterion. The new choice*

$$B = \begin{pmatrix} 0 & 1 \\ 1 & -1 \end{pmatrix}, \quad N = \begin{pmatrix} 1 & 2 \\ 0 & 1 \end{pmatrix}, \quad c_B = (1 \quad 1), \quad c_N = (3 \quad 9),$$

leads us to find

$$x_B = \begin{pmatrix} 0 \\ 2 \end{pmatrix},$$

and the new extremal vector is again $(0, 2, 0, 0)$ with associated cost 2. The vectors z and r are

$$z = (2 \quad 1), \quad r = (3 \quad 9) - (2 \quad 1)\begin{pmatrix} 1 & 2 \\ 0 & 1 \end{pmatrix} = (1 \quad 4).$$

Since all components of r are nonnegative, we have ended our search as we had anticipated: The minimum cost is 2 and it is taken on at the vector $(0, 2, 0, 0)$.

Example 2.8   *(No solution)*

$$\text{Minimize} \quad -3x_1 + x_2 + 9x_3 + x_4$$

subject to

$$x_1 - 2x_3 - x_4 = -2,$$
$$x_2 + x_3 - x_4 = 2,$$
$$x_i \geq 0.$$

In this case,

$$A = \begin{pmatrix} 1 & 0 & -2 & -1 \\ 0 & 1 & 1 & -1 \end{pmatrix}, \quad b = \begin{pmatrix} -2 \\ 2 \end{pmatrix}, \quad c = (-3 \quad 1 \quad 9 \quad 1).$$

1. Initialization. If we were to choose

$$B = \begin{pmatrix} 1 & 0 \\ 0 & 1 \end{pmatrix}, \quad N = \begin{pmatrix} -2 & -1 \\ 1 & -1 \end{pmatrix}, \quad c_B = (-3 \quad 1), \quad c_N = (9 \quad 1),$$

then we would obtain

$$x_B = b = \begin{pmatrix} -2 \\ 2 \end{pmatrix}$$

which is not a feasible vector, since there is one negative coordinate. Let us take instead (second and fourth columns of A)

$$B = \begin{pmatrix} 0 & -1 \\ 1 & -1 \end{pmatrix}, \quad N = \begin{pmatrix} 1 & -2 \\ 0 & 1 \end{pmatrix}, \quad c_B = (1 \quad 1), \quad c_N = (-3 \quad 9).$$

2. Checking the stopping criterion. It is easy to find

$$x_B = \begin{pmatrix} 4 \\ 2 \end{pmatrix}$$

such that the initial vertex is $(0, 4, 0, 2)$ with cost 6. On the other hand,

$$z = (-2 \quad 1), \quad r = (-3 \quad 9) - (-2 \quad 1) \begin{pmatrix} 1 & -2 \\ 0 & 1 \end{pmatrix} = (-1 \quad 4).$$

Since not all components of r are nonnegative, we must go through the iterative process in the simplex method.

3. Iterative step. Choose x_1 as the entering variable, since it is the one associated with the negative component of r. Moreover,

$$w = \begin{pmatrix} -1 \\ -1 \end{pmatrix}, \quad \frac{x_B}{w} = \{-4, -2\}.$$

In this situation, we have no choice for the leaving variable: no positive denominator. This means that the proposed LPP does not admit an optimal solution, i.e., the cost can be lowered indefinitely. This can be easily checked by considering the feasible vectors

$$\begin{pmatrix} t - 2 \\ t + 2 \\ 0 \\ t \end{pmatrix}, \quad t > 0.$$

The cost associated with such points is $8 - t$, which can clearly be sent to $-\infty$ by taking t sufficiently large.

Example 2.9   *(Multiple solution)*

$$\text{Minimize} \quad 3x_1 + 2x_2 + 8x_3 + x_4$$

subject to

$$x_1 - 2x_3 - x_4 = -2,$$
$$x_2 + x_3 - x_4 = 2,$$
$$x_i \geq 0.$$

In order to argue that there are infinitely many optimal solutions for this LPP, we will use the equality constraints to "solve" for x_1 and x_2 and take these expressions back into the objective function. Namely,

$$x_1 = 2x_3 + x_4 - 2 \geq 0, \quad x_2 - x_3 + x_4 + 2 \geq 0,$$

and the cost function becomes

$$6(2x_3 + x_4) - 2.$$

Since the first constraint reads

$$2x_3 + x_4 \geq 2,$$

it is clear that the minimum value of the cost will be achieved when

$$2x_3 + x_4 = 2.$$

We have the two basic solutions $(0, 1, 1, 0)$ and $(0, 4, 0, 2)$. Any convex combination of these two will also be an optimal solution

$$t(0, 1, 1, 0) + (1 - t)(0, 4, 0, 2), \quad t \in [0, 1].$$

We believe that it is elementary to understand the way in which the simplex method works after several examples. Computations can, however, be organized in tables (tableaux) to facilitate the whole process without having to explicitly write down the different steps as we have done in the previous examples. We will treat some of these practical issues in a subsequent section.

3. DUALITY

Duality is a concept that intimately links the two following LPP:

$$\text{Minimize} \quad cx \quad \text{subject to} \quad Ax \geq b, \quad x \geq 0;$$
$$\text{Maximize} \quad yb \quad \text{subject to} \quad yA \leq c, \quad y \geq 0.$$

We will identify the first problem as the primal, and the second one as its associated dual. Notice how the same elements, the matrix A and the vectors b and c, determine both problems.

Definition 2.10   (Dual problem) The dual problem of the LPP

$$\text{Minimize} \quad cx \quad \text{subject to} \quad Ax \geq b, \quad x \geq 0$$

is the LPP

$$\text{Maximize} \quad yb \quad \text{subject to} \quad yA \leq c, \quad y \geq 0.$$

Although this format is not the standard one we have utilized in our discussion of the simplex method, it allows us to see in a more transparent fashion that the dual of the dual is the primal. This is, in fact, very easy to check by transforming minima to maxima and reversing inequalities by appropriately using minus signs (this is left to the reader).

If the primal problem is formulated in the standard form

$$\text{Minimize} \quad cx \quad \text{under} \quad Ax = b, \quad x \geq 0,$$

what is its dual version? Answering this question is an elementary exercise that involves writing an LPP in standard form in the above format, applying duality, and then trying to simplify the final form of the dual. As a matter of fact, all we have to do is put

$$Ax = b \quad \text{is equivalent to} \quad Ax \geq b, \quad -Ax \geq -b,$$

so that if we write
$$\overline{A} = (A \quad -A), \quad \overline{b} = (b \quad -b),$$

our initial LPP is

$$\text{Minimize} \quad cx \quad \text{under} \quad \overline{A}x \geq \overline{b}, \quad x \geq 0.$$

Therefore, its dual will have the form

$$\text{Maximize} \quad \overline{y}\overline{b} \quad \text{subject to} \quad \overline{y}\overline{A} \leq c, \quad \overline{y} \geq 0.$$

If we now try to simplify the formulation of this problem by setting

$$\overline{y} = (y^{(1)} \quad y^{(2)}),$$

we arrive at

$$\text{Maximize} \quad (y^{(1)} - y^{(2)})b \quad \text{under} \quad (y^{(1)} - y^{(2)})A \leq c, \quad \overline{y} \geq 0,$$

and letting $y = y^{(1)} - y^{(2)}$, there is no restriction on the sign of y, and we have

$$\text{Maximize} \quad yb \quad \text{under} \quad yA \leq c,$$

which is the form of the dual when the primal is given in the standard form.

Lemma 2.11   *(Dual problem in standard form) If the primal problem is*

$$\text{Minimize} \quad cx \quad \text{under} \quad Ax = b, \quad x \geq 0,$$

its dual is

$$\text{Maximize} \quad yb \quad \text{under} \quad yA \leq c.$$

Before proceeding in analyzing duality more formally, it may be worthwhile to motivate this analysis by providing an interpretation of the meaning of the relationship between a primal and its dual. Indeed, it is interesting to realize that they are different, but equivalent, ways of looking at the same underlying problem. We are going to emphasize this point by describing a typical LPP related to networks. The practical implications of duality are often tied to the underlying problem behind a formal LPP. We restrict attention here to formally checking the equivalence of a primal and its dual without paying much attention to other implications. A full analysis and understanding of these would be required in realistic situations.

Example 2.12   *We wish to send a certain product from node A to node D in the simplified network of Figure 2.5.*

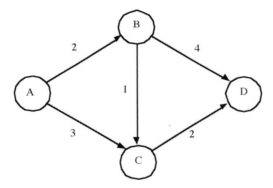

Figure 2.5. A simple network.

As you can see, we have five possible channels with associated costs given in the same figure. If we use variables x_{PQ} to denote the fraction of the product transferred through channel PQ, we must minimize the total cost

$$2x_{AB} + 3x_{AC} + x_{BC} + 4x_{BD} + 2x_{CD}$$

subject to the restrictions

$$x_{AB} = x_{BC} + x_{BD} \quad \text{(no part of the product is lost at node B)},$$
$$x_{AC} + x_{BC} = x_{CD} \quad \text{(no part of the product is lost at node C)},$$
$$x_{BD} + x_{CD} = 1 \quad \text{(the total amount of the product reaches node D)},$$
$$x_{AB}, x_{AC}, x_{BC}, x_{BD}, x_{CD} \geq 0.$$

Notice that as a consequence of these restrictions, it is easy to obtain

$$x_{AB} + x_{AC} = 1,$$

so that the total amount of the product departs from node A. This is the primal formulation of the problem.

We can also think in terms of prices per unit of product at the different nodes of the network y_A, y_B, y_C, y_D and consider the differences between these prices at nodes as the profit when that particular channel is used. In this situation we are seeking the maximum for $y_D - y_A$, the profit in transferring the good from A to D. The profits for the five channels will be

$$y_B - y_A, \quad y_C - y_A, \quad y_C - y_B, \quad y_D - y_B, \quad y_D - y_C.$$

If we take as a normalization rule $y_A = 0$, then we must demand that these profits not exceed the prices for the use of each channel:

$$y_B - y_A = y_B \leq 2, \quad y_C - y_A = y_C \leq 3, \quad y_C - y_B \leq 1,$$
$$y_D - y_B \leq 4, \quad y_D - y_C \leq 2.$$

This would be the dual formulation of the problem.

We somehow suspect that these two problems must be equivalent and that their optimal solutions must be related to each other. Indeed, this is the case. The connection is precisely the duality link. With the elements

$$c = (2 \quad 3 \quad 1 \quad 4 \quad 2), \quad b = \begin{pmatrix} 0 \\ 0 \\ 1 \end{pmatrix},$$

$$A = \begin{pmatrix} 1 & 0 & -1 & -1 & 0 \\ 0 & 1 & 1 & 0 & -1 \\ 0 & 0 & 0 & 1 & 1 \end{pmatrix},$$

these two problems can be formulated in compact form as

$$\text{Minimize} \quad cx \quad \text{under} \quad Ax = b, \quad x \geq 0,$$
$$\text{Maximize} \quad yb \quad \text{under} \quad yA \leq c,$$

where

$$x = (x_{AB}, x_{AC}, x_{BC}, x_{BD}, x_{CD}), \quad y = (y_B, y_C, y_D).$$

This is precisely the form of a primal and its dual.

Next we briefly describe in two steps the relationship between the solutions of a primal problem (P) and its dual (D), where we assume that (P) is given in standard form.

Lemma 2.13   *(Weak duality) If x and y are feasible for (P) and (D), respectively, then*

$$yb \leq cx.$$

Moreover, if equality holds,

$$yb = cx,$$

then x is an optimal solution for (P) and y for (D).

The proof is rather simple. Indeed, from

$$Ax = b, \quad x \geq 0, \quad yA \leq c,$$

we have

$$yb = yAx \leq cx.$$

In particular, we have

$$\max\{yb : yA \leq c\} \leq \min\{cx : Ax = b, x \geq 0\}.$$

If $yb = cx$, this number must be at the same time the previous maximum and minimum, and this in turn implies that y is optimal for (D) and x for (P).

This result also informs us that degenerate cases, when either the maximum for the dual is $+\infty$ or the minimum for the primal is $-\infty$, occur when the other problem does not have feasible vectors.

The full duality theorem follows.

Theorem 2.14   *(Duality theorem) Either both problems (P) and (D) are solvable simultaneously, or one of the two is degenerate in the sense that it does not admit feasible vectors.*

What this statement amounts to is that if x is optimal for (P), then there exists an optimal vector y for (D), and the common value $yb = cx$ is at the same time the minimum for (P) and the maximum for (D). Conversely, if y

is optimal for (D), there exists an optimal vector x for (P) with the common value $yb = cx$ being at the same time the minimum and the maximum.

The proof relies on our previous discussion of the simplex method. If x is optimal for (P), then

$$x = (\,x_B \quad 0\,), \quad x_B = B^{-1}b, \quad cx = c_B B^{-1}b,$$
$$r = c_N - c_B B^{-1}N \geq 0 \text{ (stopping criterion)},$$

where $c = (\,c_B \quad c_N\,)$. If we examine $y = c_B B^{-1}$, it turns out that

$$yA = c_B B^{-1}(\,B \quad N\,) = (\,c_B \quad c_B B^{-1}N\,) \leq (\,c_B \quad c_N\,) = c,$$

so y is admissible for (D). On the other hand, y is such that

$$cx = c_B B^{-1}b = yb.$$

By the weak duality principle, x and y must be optimal.

The fact that the dual of the dual is the primal lets us pass from an optimal solution for (D) to an optimal solution for (P).

Notice how an optimal solution for the dual has been obtained from an optimal solution of the primal: If $x = (\,x_B \quad 0\,)$ is optimal for (P) with

$$x_B = B^{-1}b, \quad c = (\,c_B \quad c_N\,),$$

then $y = c_B B^{-1}$ is optimal for (D). We will return to this observation later.

It is important to stress the information on the primal provided by the optimal solution of the dual. One such interpretation comes directly from the duality theorem, and refers to how changes to the vector b affect the optimal value of the primal. This is an issue of great practical importance, since we are also interested in assessing how good changes in the independent term b are. Constraints involving b are typically related to restrictions such as production capacities and total investments, and we would like to know whether making such changes would pay off. If $M(b)$ stands for the dependence of the value of the minimum of an LPP problem on b, we are seeking the partial derivatives ∇M. These are called sensitivity parameters, shadow prices, and dual prices as well. By the duality theorem,

$$M(b) = yb,$$

where y is the optimal solution of the dual. Intuitively,

$$\nabla M(b) = y$$

and this is usually interpreted by saying that the dual solution provides the rate of change of the optimal value for the primal when the vector b of restrictions changes. It is therefore as important to know the optimal solution of an LPP as the optimal solution of the dual. To be rigorous, the above computation of ∇M has not been justified, since the optimal solution for the dual depends on b. But since the result is essentially correct and plausible, we will not insist on this point.

4. SOME PRACTICAL ISSUES

In the preceding sections, we have been concerned with the understanding of the structure of an LPP and the standard mechanism to solve it by the simplex method. There are, however, a number of issues of some practical importance. We will treat in this section three such topics:

1. how to initialize the simplex method from a practical viewpoint;
2. how to organize computations in an efficient manner through tables;
3. how the optimal solution of the dual can be rapidly found from the solution of the primal.

The significance of such issues is of relative value, since as soon as the number of variables involved in an LPP exceeds a few, software packages must be employed to find optimal solutions in reasonable periods of time.

In our discussion of the simplex method, we have not indicated how to find a first feasible choice for the matrix B. This amounts to selecting m columns among the n columns of A such that the solution of the linear system $Bx = b$ is such that $x \geq 0$. In some cases, doing this directly may be a rather tedious task. What we would like to do is to describe a more-or-less efficient mechanism that may lead us to find such a feasible submatrix B without going over an exhaustive enumeration of all possibilities, which would be, after all, solving the LPP by brute force. We will describe two different ways of finding such an initialization.

The first one relies on an auxiliary LPP, with trivial initialization, whose optimal solution will tell us how to choose the initial feasible submatrix B. The auxiliary problem is

$$\text{Minimize} \quad \sum_i \tilde{x}_i$$

subject to

$$(\tilde{A} \quad \mathbf{1})\, X = \tilde{b}, \quad X \geq 0,$$

where $X = (x \quad \tilde{x})$, and $\tilde{A}$ and $\tilde{b}$ are such that the system $Ax = b$ is equivalent to $\tilde{A}x = \tilde{b}$ but $\tilde{b} \geq 0$. This can be simply done by multiplying by -1 those constraints associated with negative components of b. Notice that a valid feasible extremal vector for this problem is $(0 \quad \tilde{b})$.

We claim that if the initial LPP admits an optimal solution x, then the minimum for the auxiliary problem is 0, and it is attained at $X = (x \quad 0)$. This is very easy to check and left to the reader as an instructive exercise. Consequently, if we solve this auxiliary LPP with initial vertex $(0 \quad \tilde{b})$ by the simplex method, the optimal solution found will be of the form $X = (x \quad 0)$, where x will have at most m nonvanishing components. Observe that the auxiliary problem has the same value for m. The positive components of this vector x will indicate which columns must be chosen for a feasible starting point for our LPP. If the number of such positive components is strictly less than the number of columns we should select, the remaining columns can be taken arbitrarily, as long as they remain a linearly independent set of vectors. In order to clarify the mechanism that leads to a feasible initialization of any LPP, let us consider the following example.

Example 2.15   *We are interested in finding a valid initialization for the LPP*

$$\text{Minimize} \quad 2x_1 + 3x_2 + x_3$$

subject to

$$x_1 + x_2 + 2x_3 + x_4 = 500,$$
$$x_1 + x_2 + x_3 - x_4 = 500,$$
$$x_1 + 2x_2 + 2x_3 = 600,$$
$$x_i \geq 0.$$

In this particular example,

$$A = \begin{pmatrix} 1 & 1 & 2 & 1 \\ 1 & 2 & 1 & -1 \\ 1 & 2 & 2 & 0 \end{pmatrix}, \quad b = \begin{pmatrix} 500 \\ 500 \\ 600 \end{pmatrix}, \quad c = (2 \quad 3 \quad 1 \quad 0).$$

The issue here is how to choose the initial matrix B in order to initialize the simplex method. In this simple case we have four possibilities that correspond to choosing three different columns from a set of four. We could certainly go through all these possibilities and choose the first basic solution with non-negative coordinates. As we have argued before, this amounts to solving the LPP itself by an exhaustive analysis of all basic solutions. When the dimension of the problem is large, this enumerative approach is not admissible, and the mechanism described to initialize the simplex method becomes interesting. In our example, this scheme would amount to considering the auxiliary LPP associated with the data

$$\tilde{A} = \begin{pmatrix} 1 & 1 & 2 & 1 & 1 & 0 & 0 \\ 1 & 2 & 1 & -1 & 0 & 1 & 0 \\ 1 & 2 & 2 & 0 & 0 & 0 & 1 \end{pmatrix}, \quad \tilde{b} = \begin{pmatrix} 500 \\ 500 \\ 600 \end{pmatrix},$$
$$c = \begin{pmatrix} 0 & 0 & 0 & 0 & 1 & 1 & 1 \end{pmatrix}.$$

The initialization for the simplex method to solve this problem is to choose the matrix B as the identity matrix for the last three columns of $\tilde{A}$. The nonvanishing components of an optimal solution for this problem, found by applying the simplex method, will indicate an initialization for our original problem. In this case, by applying the simplex method we obtain the optimal solution

$$(200, 100, 100, 0, 0, 0, 0),$$

and this indicates that the initial matrix B, made up of the three first columns of A, is a valid initialization for the simplex method for our original LPP.

The second approach to the initialization issue does not require us to consider an additional, auxiliary, LPP. It is based instead on transforming the given LPP into a new, equivalent, LPP for which the initialization is trivial. Rather than stating a formal result, we will intuitively discuss this transformation. Consider a typical LPP,

$$\text{Minimize} \quad cx \quad \text{under} \quad Ax = b, \quad x \geq 0,$$

where $b \geq 0$ (multiplying by -1 those equations corresponding to negative components of b). Let us introduce new variables $y \in \mathbf{R}^m$ and study the LPP

$$\text{Minimize} \quad cx + dy \quad \text{under} \quad Ax + y = b, \quad x \geq 0, \quad y \geq 0,$$

where the vector $d \in \mathbf{R}^m$ is assumed to have very large unspecified components. The whole point is that $(0 \quad b)$ is a valid initialization for the second LPP, and for sufficiently large components in the vector d, the optimal solution will be of the form $(x \quad 0)$, where x is an optimal solution of our initial LPP. The first assertion is trivial. The second is plausible, since if the components of d are very large and we are seeking to minimize the sum $cx + dy$ with $y \geq 0$, we see that optimal solutions will essentially require $y = 0$, and therefore we fall back in our initial LPP. The disadvantage of this procedure is that in solving this transformed LPP, we must work symbolically with the vector d, or else we should assign a very large value for d. We will see an example later.

The computations involved in the simplex method are usually organized in the form of tables reflecting the different steps we have already described in Section 2.2. Since the simplex method proceeds by changing feasible submatrices B from the original matrix A, and renaming columns so that the first m columns correspond to those in B, it is extremely important not to get lost with such reorganization and to keep a record of the initial enumeration of columns regardless of the position they ocupy in the succesive steps. Each of these tables has the structure

$$A \qquad b$$

$$c \qquad d,$$

where d is the value indicating the cost, changed in sign, of the different basic feasible solutions the simplex method passes through. In each of these tables the following calculations must be performed succesively:

1. Choose the columns corresponding to the next feasible submatrix B, place them on the first m columns of the table, and by using the elementary transformations of linear algebra obtain the identity matrix from the columns of B (it is not enough to have an upper or lower triangular matrix). Do not forget to keep a record of the columns belonging to B.

2. Transform in the same manner the components of c (last row) so that those corresponding to the columns of B vanish.

3. If the remaining components of c are nonnegative (stopping criterion), one (the) optimal solution is found by solving the linear system $Bx = b$ with the current submatrix B and independent term b, putting zero in the components not contained in B (this is why it is important to keep track of which columns

are part of the submatrices B). If there are some negative components, we
select the entering column as the one with the least component in c.

4. Examine the ratios of b over the entering column componentwise; if there
is no positive denominator, the problem lacks an optimal solution; otherwise,
choose as leaving column the one associated with the least nonnegative ratio
among the selected ones. Go back to 1 until the stopping criterion is fulfilled
or we reach the conclusion that there does not exist an optimal solution.

Instead of insisting on clarifying these steps, which faithfully reflect those
described in Section 2.2, we propose to examine one concrete example.

Example 2.16   *We want to minimize* $-3x_1 - 5x_2$ *under the constraints* $x \geq 0$
and   tablecu method
$$3x_1 + 2x_2 + x_3 = 18, \quad x_1 + x_4 = 4, \quad x_2 + x_5 = 6.$$

The initial table for this example is

x_1	x_2	x_3	x_4	x_5	
3	2	1	0	0	18
1	0	0	1	0	4
0	1	0	0	1	6
-3	-5	0	0	0	0.

If we choose the columns or variables 3, 4, and 5 to make up the matrix
B, *we find that the vertex* $(0, 0, 18, 4, 6)$ *is feasible. If we reorganize the three
selected columns as the first three columns, we have the table*

x_3	x_4	x_5	x_1	x_2	
1	0	0	3	2	18
0	1	0	1	0	4
0	0	1	0	1	6
0	0	0	-3	-5	0.

Since in this particular case the matrix B *is already the identity, no further
computation is required on the table for this purpose. On the other hand, the
components of* c *not corresponding to columns in* B *are both negative* (-3 *and*
-5); *hence the stopping criterion does not hold, and we ought to transform the
table according to the main step of the simplex method. The entering variable
would be* x_2 *(associated with* -5 *in* c). *To determine the leaving variable, we*

examine the ratios $18/2$ and $6/1$ ($4/0$ is discarded because it has a vanishing denominator) and select the third ratio 6 as the smallest one. Accordingly, the third column of B (corresponding to x_5) is the leaving column. After these two variables are interchanged, the table looks like this:

x_3	x_4	x_2	x_1	x_5	
1	0	2	3	0	18
0	1	0	1	0	4
0	0	1	0	1	6
0	0	-5	-3	0	0.

With this new table we should obtain, by means of elementary transformations, the identity matrix in the first three columns, and the null vector on the three components of c. In this particular case, these two objectives are obtained by changing the first row to itself minus twice the third one, and replacing the fourth one by itself plus five times the third one. After these changes we arrive at

x_3	x_4	x_2	x_1	x_5	
1	0	0	3	-2	6
0	1	0	1	0	4
0	0	1	0	1	6
0	0	0	-3	5	30.

Once again we look at the nonvanishing components of c and select the least one (negative) as entering variable (x_1). To choose the leaving one, we examine the ratios $6/3$, $4/1$, and choose the smallest among the positive ones, associated, in this case, with x_3. These changes lead to

x_1	x_4	x_2	x_3	x_5	
3	0	0	1	-2	6
1	1	0	0	0	4
0	0	1	0	1	6
-3	0	0	0	5	30.

As before, we seek the identity matrix in the first three columns, and the null vector on the three components of c. The new table is

x_1	x_4	x_2	x_3	x_5	
1	0	0	$1/3$	$-2/3$	2
0	1	0	$-1/3$	$2/3$	2
0	0	1	0	1	6
0	0	0	1	3	36.

Since in this table the stopping criterion holds (all nonvanishing components of c are nonnegative), the optimal solution is found in column b. The optimal cost (changed in sign) appears in d, -36. It is important to determine the components associated with the values in b. In the last table the matrix B is formed by the three columns x_1, x_4, and x_2, and the components of b will correspond (in this order) to these variables. The variables not present in B are assigned a vanishing value. Thus the optimal solution is $(2, 6, 0, 2, 0)$ with optimal cost -36. In practice, computations proceed by transforming the tables without any further comment.

We solve another example including the discussion on initialization by the second method we have indicated before.

Example 2.17   *The problem is*   full sequence of tableau method

$$\text{Minimize} \quad 3x_1 + x_2 + 9x_3 + x_4$$

subject to

$$x_1 + 2x_3 + x_4 = 4,$$
$$x_2 + x_3 - x_4 = 2,$$
$$x_i \geq 0.$$

According to our discussion on how to set up a new equivalent optimization problem for which initialization is trivial, we consider the modified LPP

$$\text{Minimize} \quad 3x_1 + x_2 + 9x_3 + x_4 + dx_5 + dx_6$$

subject to

$$x_1 + 2x_3 + x_4 + x_5 = 4,$$
$$x_2 + x_3 - x_4 + x_6 = 2,$$
$$x_i \geq 0,$$

where d is assumed to be a very large parameter. The initial table for this problem is

x_1	x_2	x_3	x_4	x_5	x_6	
1	0	2	1	1	0	4
0	1	1	−1	0	1	2
3	1	9	1	d	d	0.

We notice that we can choose as an admissible initialization the identity matrix corresponding to the fifth and sixth columns. The second table is

x_5	x_6	x_1	x_2	x_3	x_4	
1	0	1	0	2	1	4
0	1	0	1	1	−1	2
0	0	$3-d$	$1-d$	$9-3d$	1	$-6d.$

If d is a very large positive number, the most negative coefficient in the last row will be $9 - 3d$, so that we choose x_3 as entering variable, and then x_1 as leaving variable (in this particular instance both ratios are equal, and we could equally choose x_6 as leaving variable). After rearranging columns and making computations we have

x_3	x_6	x_1	x_2	x_5	x_4	
1	0	1/2	0	1/2	1/2	2
0	1	−1/2	1	−1/2	−3/2	0
0	0	$(d-3)/2$	$1-d$	$3(d-3)/2$	$(3d-7)/2$	$-18.$

Again, having in mind that d is a very large positive number, we would choose x_2 as entering variable and x_6 as our leaving variable. After computations the table is

x_3	x_2	x_1	x_6	x_5	x_4	
1	0	1/2	0	1/2	1/2	2
0	1	−1/2	1	−1/2	−3/2	0
0	0	−1	$d-1$	$d-4$	−2	$-18.$

Our next entering variable is x_4, and x_3 our leaving variable. We obtain

x_4	x_2	x_1	x_6	x_5	x_3	
1	0	1	0	1	2	4
0	1	1	1	1	3	6
0	0	1	$d-1$	$d-2$	4	$-10.$

Since all nonvanishing coefficients in the last row are positive, we have already found the optimal solution, which is given by the last column 4, 6 for x_4 and x_2, respectively, and the remaining variables should be set to zero. The optimal cost is 10 and the optimal solution $(0, 6, 0, 4, 0, 0)$. Notice how this optimal solution has vanishing components for the auxiliary variables x_5 and x_6. The optimal solution for the original problem will be $(0, 6, 0, 4)$ with optimal cost 10.

Altenatively, we can also assign a very large numeric value to d (much larger than any of those participating in the problem, for instance $d = 100$) and solve the problem. If the final solution yields vanishing values for x_5 and x_6, we have our optimal solution. If not, the problem must be solved again with a larger value for d.

Finally, we want to stress more explicitly how the passage from the optimal solution of the primal to the optimal solution of the dual can be done in an efficient manner. In fact, this was indicated when we treated duality, but we would like to emphasize it here. Assume that the primal is

$$\text{Minimize} \quad cx \quad \text{under} \quad Ax = b, \quad x \geq 0,$$

with dual

$$\text{Maximize} \quad yb \quad \text{under} \quad yA \leq c.$$

If

$$x = (\, x_B \quad 0\,), \quad x_B = B^{-1}b,$$

is the optimal solution of the primal, the optimal solution of the dual will be

$$y = c_B B^{-1},$$

where c_B incorporates the components of c corresponding to columns selected in B. In practice, it is a matter of solving the linear system

$$c_B = yB,$$

where B and c_B include the columns for the optimal solution. Indeed, these columns are associated with the inequalities that must be converted to equalities to find the optimal solution of the dual. One example will clarify this last sentence.

Example 2.18   We want to maximize the function $18y_1 + 4y_2 + 6y_3$ under the constraints

$$3y_1 + y_2 \leq -3, \quad 2y_1 + y_3 \leq -5, \quad y \leq 0.$$

In matrix form, these restrictions can be written as

$$\begin{pmatrix} 3 & 1 & 0 \\ 2 & 0 & 1 \\ 1 & 0 & 0 \\ 0 & 1 & 0 \\ 0 & 0 & 1 \end{pmatrix} \begin{pmatrix} y_1 \\ y_2 \\ y_3 \end{pmatrix} \leq \begin{pmatrix} -3 \\ -5 \\ 0 \\ 0 \\ 0 \end{pmatrix}.$$

We have used the y-variable to suggest that this problem may be diretly understood as the dual of a certain primal LPP. It is true that we can solve it directly by transforming it to standard form and applying the simplex method. But this process requires more labor than if we treat it as a dual problem. In fact, its associated primal is

$$\text{Minimize} \quad -3x_1 - 5x_2$$

subject to

$$3x_1 + 2x_2 + x_3 = 18, \quad x_1 + x_4 = 4,$$
$$x_2 + x_5 = 6, \quad x \geq 0.$$

This problem has already been solved. Its optimal solution is the vector

$$(2, 6, 0, 2, 0).$$

The nonvanishing components of this vector indicate that the matrix B includes the first, second, and fourth columns of A. If we had only two (or fewer) nonvanishing components, the third column could be arbitrarily chosen as long as the resulting matrix were nonsingular. This information suffices to solve the dual, since its optimal solution can be found through the solution of the linear system

$$
\begin{pmatrix} 3 & 1 & 0 \\ 2 & 0 & 1 \\ 0 & 1 & 0 \end{pmatrix} \begin{pmatrix} y_1 \\ y_2 \\ y_3 \end{pmatrix} = \begin{pmatrix} -3 \\ -5 \\ 0 \end{pmatrix},
$$

obtained by transforming into equalities the first, second, and fourth inequalities of the dual. The optimal solution is thus $(-1, 0, -3)$ and the optimal cost -36.

5. INTEGER PROGRAMMING

Engineering applications of linear programming often require variables to take on integer values rather than real ones. In some cases, neglecting this restriction would provide a reasonable approximation. In others, it is crucial to pay close attention to this constraint. In such cases, in addition to the typical linear constraints

$$Ax = b, \quad x \geq 0,$$

we must force some (or all) variables to take on (nonnegative) integer values $x_i \in \mathbf{Z}$. This new constraint will be dictated by the nature of the problem we are interested in. We therefore face the LPP, which we will identify as $(\tilde{P})$,

$$\text{Minimize} \quad cx \quad \text{under} \quad Ax = b, \quad x \geq 0,$$
$$x_i \in \mathbf{Z}, \quad i \in I \subset N = \{1, 2, \ldots, n\},$$

where the subset of indices I is known. Reasonably enough, we will care first about the underlying LPP (P) without the integer constraint

$$\text{Minimize} \quad cx \quad \text{under} \quad Ax = b, \quad x \geq 0.$$

Assume for a moment that $x^{(0)}$, the optimal solution for (P), satisfies the integrality requirement $x_i^{(0)} \in \mathbf{Z}$. In this case, it is evident that we have found the (or one) optimal solution for $(\tilde{P})$. Most likely, however, we will not be so lucky, and such an optimal solution will not verify the complete integrality condition. How will we proceed in such a situation? The scheme to follow is called the "branch and bound method," and it consists in generating a sequence of sub-problems, solving them, and analyzing and comparing the different solutions until we reach a feasible optimal solution for our original problem.

The basic idea behind decomposing a problem into two disjoint subproblems ("branching") is the following. Assume, by recursion, that we have one LPP as a result of previous steps and we have not yet found a feasible vector for our initial problem. We find its optimal solution $x^{(0)}$. Obviously, if the problem is infeasible, it is discarded altogether. Two situations may occur:

1. If $x^{(0)}$ satisfies the integrality constraints, it becomes our provisional optimal solution, and we discard the subproblem;
2. if $x^{(0)}$ does not satisfy all of the integrality constraints, choose one variable

$$x_i^{(0)} \in (k, k+1), \quad k \in \mathbf{Z}, \quad i \in I,$$

and add to the collection of subproblems to be analyzed the two disjoint subproblems ("branching") obtained by adding to the constraints of the problem the further constraint $x_i \leq k$ in one case, and $x_i \geq k+1$ in the other. Notice how the feasible sets for these two subproblems are disjoint, and their union is the complete feasible set of the LPP they come from. See Figure 2.6.

Once we have found a feasible provisional optimal solution x^*, and we have to analyze one subproblem previously generated by the branching procedure, the discussion would be as follows:

1. If
$$cx^* \leq cx^{(0)},$$

we discard this LPP altogether, since it cannot improve the optimal solution we have already found, and choose another subproblem;

2. if
$$cx^* > cx^{(0)},$$

and $x^{(0)}$ satisfies the integrality requirement, change the provisional optimal solution to $x^{(0)}$, discard the corresponding problem, and analyze another subproblem;

3. if
$$cx^* > cx^{(0)},$$

and $x^{(0)}$ does not satisfy the complete integrality constraint, proceed to branch this problem as indicated above.

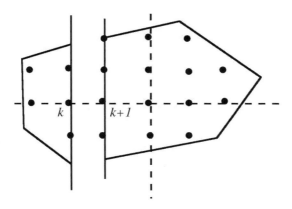

Figure 2.6. Branching of a domain.

In this way, we are ensuring that the optimal solution will be found by this exhaustive process. Again, whenever subproblems are infeasible due to lack of feasible vectors, they are eliminated.

After all of the generated subproblems have been analyzed, the provisional optimal solution indeed becomes the optimal solution of our initial problem. This is always a finite process.

Example 2.19   *We want to solve the problem* $(\tilde{P})$

$$\text{Minimize} \quad 3x_2 + 2x_3$$

under

$$2x_1 + 2x_2 - 4x_3 = 5, \quad 4x_2 + 2x_3 \leq 3,$$
$$x_i \geq 0, \quad x_1, x_3 \in \mathbf{Z}.$$

The underlying LPP is

$$\text{Minimize} \quad 3x_2 + 2x_3 \quad \text{under}$$
$$2x_1 + 2x_2 - 4x_3 = 5, \quad 4x_2 + 2x_3 \leq 3, \quad x_i \geq 0,$$

whose optimal solution is $(5/2, 0, 0)$. Since x_1 is not an integer, we must proceed to "branch" this problem. We still do not have a feasible provisional solution. The two subproblems are

$$\text{Minimize} \quad 3x_2 + 2x_3 \quad \text{under}$$
$$2x_1 + 2x_2 - 4x_3 = 5, \quad 4x_2 + 2x_3 \le 3,$$
$$x_1 \le 2, \quad x_i \ge 0,$$

and

$$\text{Minimize} \quad 3x_2 + 2x_3 \quad \text{under}$$
$$2x_1 + 2x_2 - 4x_3 = 5, \quad 4x_2 + 2x_3 \le 3,$$
$$x_1 \ge 3, \quad x_i \ge 0.$$

Their respective optimal solutions are

$$\left(2, \frac{1}{2}, 0\right), \quad \text{cost} = \frac{3}{2},$$
$$\left(3, 0, \frac{1}{4}\right), \quad \text{cost} = \frac{1}{2}.$$

The first of these respects the integrality requirement, and therefore it is taken as our provisional solution. Since the cost of the second is smaller than the cost for the provisional solution, this subproblem must be considered, since it could contain a better solution. On the other hand, since the second solution does not respect the integrality restriction, we must branch this subproblem and obtain

$$\text{Minimize} \quad 3x_2 + 2x_3 \quad \text{under}$$
$$2x_1 + 2x_2 - 4x_3 = 5, \quad 4x_2 + 2x_3 \le 3,$$
$$x_1 \ge 3, \quad x_3 \le 0, \quad x_i \ge 0,$$

and

$$\text{Minimize} \quad 3x_2 + 2x_3 \quad \text{under}$$
$$2x_1 + 2x_2 - 4x_3 = 5, \quad 4x_2 + 2x_3 \le 3,$$
$$x_1 \ge 3, \quad x_3 \ge 1, \quad x_i \ge 0.$$

The first one is infeasible since $x_3 = 0$, and from the first constraint we get $2x_1 + 2x_2 = 5$. This together with $x_1 \ge 3$ and $x_2 \ge 0$ is impossible. This *subproblem is thus discarded. The optimal solution for the second is $(9/2, 0, 1)$ with cost 2. Since this cost is greater than the one for the provisional solution, we eliminate this subproblem without changing the optimal solution. Since there are no more subproblems to analyze, the provisional optimal solution is proclaimed as the optimal solution for the initial problem*

$$\left(2, \frac{1}{2}, 0\right), \quad cost = \frac{3}{2}.$$

Notice how it differs from the optimal solution without the integrality requirement.

In practice, it is not an easy task to decide on the variable to be branched in such a way that the whole process turns out to be as short and efficient as possible. There are no fixed rules to determine the most efficient choice in any situation. Any available a priori information on the problem may dictate that one should follow one particular path over other possible alternatives.

6. EXERCISES

1. Draw in the plane the region determined by the inequalities

$$x_2 \geq 0, \quad 0 \leq x_1 \leq 3, \quad -x_1 + x_2 \leq 1, \quad x_1 + x_2 \leq 4.$$

Find the point(s) where the following functions attain their maximum and minimum values:

$$2x_1 + x_2, \quad x_1 + x_2, \quad x_1 + 2x_2.$$

2. Solve graphically the next two problems:

$$\text{Maximize} \quad 2x_1 + 6x_2$$

subject to

$$-x_1 + x_2 \leq 1, \quad 2x_1 + x_2 \leq 2, \quad x_1 \geq 0, x_2 \geq 0;$$

$$\text{Minimize} \quad -3x_1 + 2x_2$$

subject to

$$x_1 + x_2 \leq 5, \quad 0 \leq x_1 \leq 4, \quad 1 \leq x_2 \leq 6.$$

3. Determine the values of the parameter d such that the feasible set determined by

$$x_1 + x_2 + x_3 \leq d, \quad x_1 + x_2 - x_3 = 1, \quad 2x_3 \geq d,$$

is empty.

4. Determine the vectors where the linear function $2x_1 + 3x_2 + x_3$ takes on its maximum under the constraints

$$x_1 \geq 0, \quad x_2 \geq 0, \quad x_3 \geq 0,$$
$$x_1 + x_2 + 2x_3 \leq 200,$$
$$3x_1 + 2x_2 + x_2 \leq 500,$$
$$x_1 + 2x_2 + 2x_3 \leq 300.$$

5. The maximum value of the function $3x_1 + 2x_2 - 2x_3$ is sought subject to the constraints

$$4x_1 + 2x_2 + 2x_3 \leq 20,$$
$$2x_1 + 2x_2 + 4x_3 \geq 6,$$
$$x_1 \geq 0, \quad x_2 \geq 0,$$

but the sign of x_3 is not restricted. Find the optimal solution(s).

6. Determine the maximum value of $18x_1 + 4x_2 + 6x_3$ under the constraints

$$3x_1 + x_2 \leq -3, \quad 2x_1 + x_3 \leq -5, \quad x_1 \leq 0, x_2 \leq 0, x_3 \leq 0,$$

by looking at the dual problem.

7. Consider the following primal problem:

$$\text{Maximize} \quad 1.1x_1 + 1.2x_2 + x_3$$

subject to

$$2x_1 + 2x_2 + 2x_3 \leq 10, \quad x_1 + 3x_2 + x_3 \leq 10, \quad 4x_1 + x_2 + x_3 \leq 10,$$
$$3x_1 + x_2 + 3x_3 \leq 10, \quad x_1 + 2x_2 + 3x_3 \leq 10, \quad 3x_1 + 2x_2 + x_3 \leq 10,$$
$$x_1 \geq 0, \quad x_2 \geq 0, \quad x_3 \geq 0.$$

Formulate the dual and solve it.

8. Find explicitly the optimal value of the LPP

$$\text{Minimize} \quad x_1 + x_2 + x_3$$

subject to

$$x_1 + 2x_2 + 3x_3 = b_1, \quad x_1 - x_2 - x_3 = b_2,$$

in terms of b_1 and b_2. Find the optimal solution of the dual problem and check its relationship to the gradient of the optimal value of the primal with respect to b_1 and b_2.

9. For the problem of the tiling elements of Chapter 1, compute the amounts of each model that can be sent in order to maximize benefits.

10. Solve the exercise of the spring system of Chapter 1 for the following data set:
    1. location of fixed nodes: $(1, 0)$, $(0, 1)$, $(-1, 0)$, $(0, -1)$;
    2. $k = 1$;
    3. $F = (1, 1)$.

11. Solve the transportation problem of Chapter 1 for the following data set:
    1. $n = 3$, $m = 2$;
    2. $u_1 = 2$, $u_2 = 2$, $u_3 = 3$, $v_1 = 5$, $v_2 = 2$;
    3. $c_{11} = 2$, $c_{12} = 1$, $c_{21} = 3$, $c_{22} = 1$, $c_{31} = 2$, $c_{32} = 3$.

12. Try to describe the best solution to the investment exercise of Chapter 1 (Exercise 1).

13. Solve the problem of the scaffolding system proposed in Chapter 1, where loads x_1 and x_2 are applied exactly at the midpoints of beams CD and EF, respectively.

14. Although there are many software packages to solve LPP of large dimension, it is not especially difficult to design a program to implement the simplex method. Do so in some language (C, Fortran, Maple, Mathematica, Matlab, etc) and use it to solve the following problems.
    1. Maximize $x_1 + x_2 - x_6$ subject to
       $$x_1 \geq 0, \ x_2 \geq 0, \ x_3 \geq 0, \ x_4 \geq 0, \ x_5 \geq 0, \ x_6 \geq 0,$$
       $$x_1 + x_2 + x_3 + x_4 + x_5 + x_6 = 1.$$
    2. Maximize $x_1 - x_2 + x_3 - x_4 + x_5 - x_6$ under the constraints
       $$x_1 \leq 0, \ x_2 \leq 0, \ x_3 \leq 0, \ x_4 \leq 0, \ x_5 \leq 0, \ x_6 \leq 0,$$
       $$x_1 + x_2 + x_3 + x_4 + x_5 + x_6 \geq -1.$$

3. Maximize $x_1 + 2x_2 + 3x_3 + 4x_4 + 5x_5 + 6x_6$ subject to

$$x_1 \geq 0, \; x_2 \geq 0, \; x_3 \geq 0, \; x_4 \geq 0, \; x_5 \geq 0, \; x_6 \geq 0,$$
$$6x_1 + 5x_2 + 4x_3 + 3x_4 + 2x_5 + x_6 \leq 1, \; 6x_1 + x_2 + 5x_3 + 2x_4 + 4x_5 + 3x_6 \leq 1.$$

4. Maximize $x_1 - x_2 + x_3 - x_4 + x_5 - x_6 + x_7 - x_8 + x_9 - x_{10}$ under the constraints

$$-1 \leq x_1 + x_2 \leq 1, \quad -1 \leq x_1 + x_2 + x_3 \leq 1,$$
$$-1 \leq x_2 + x_3 + x_4 \leq 1, \quad -1 \leq x_3 + x_4 + x_5 \leq 1,$$
$$-1 \leq x_4 + x_5 + x_6 \leq 1, \quad -1 \leq x_5 + x_6 + x_7 \leq 1,$$
$$-1 \leq x_6 + x_7 + x_8 \leq 1, \quad -1 \leq x_7 + x_8 + x_9 \leq 1,$$
$$-1 \leq x_8 + x_9 + x_{10} \leq 1, \quad -1 \leq x_9 + x_{10} \leq 1.$$

15. Some nonlinear functions may be treated in the context of LPP. Try to formulate and solve the following problem:

$$\text{Minimize} \quad |x_1| - x_2$$

subject to

$$x_1 + |x_2| \leq 1, \quad 2|x_1| - |x_2| \leq 2.$$

(Hint: The function $|\cdot|$ can be modeled in a linear fashion by decomposing it into the sum of two independent nonnegative variables, just as a variable that is unrestricted in sign is the difference of two such independent variables.)

16. Consider the simple LPP

$$\text{Maximize} \quad x_1 + 2x_2$$

subject to $x_1 + x_2 \leq 1, \, 0 \leq x_1 \leq 1, \, 0 \leq x_2$. Check that the simplex method enters into a cyclic infinite process by choosing as initialization the matrix corresponding to the variables x_1, x_2. Notice how the inequality $x_1 \leq 1$ is redundant with $x_1 + x_2 \leq 1, \, 0 \leq x_1, x_2$. Find out whether by eliminating such an inequality, the simplex method avoids the cyclic process.

Chapter 3

Nonlinear Programming

1. MODEL PROBLEM

The problem we will be concerned with in this chapter has a similar structure to that of an LPP. We would like to learn how to

$$\text{Minimize} \quad C(x) \quad \text{under} \quad A(x) \leq 0.$$

In the situation of an LPP both the cost functional C and the functions determining admissibility A were linear. If either C or some of the components of A are nonlinear, the previous problem is said to be a nonlinear programming problem (NLPP). As our readers may easily infer, these problems are considerably more complex than their linear counterparts. We assume that all functions are smooth, unless otherwise explicitly stated.

Although we have written the constraints in the form of inequalities, they can be formulated as equalities and inequalities, as has been indicated in the previous chapter, where we emphasized that multiplying by -1 reverses the direction of an inequality, and that an equality is equivalent to two inequalities. Because constraints in the form of equalities and inequalities play a different role in NLPP, they are typically distinguished by different names, so that throughout this chapter, we will stick to the following general form of an NLPP:

Definition 3.1   *The standard form of an NLPP is*

$$\text{Minimize}\quad f(x)\quad \text{subject to}\quad g(x) \leq 0, \quad h(x) = 0,$$

where $x \in \mathbf{R}^n$.

A first important question is related to the existence of optimal solutions for this problem. We already know that even LPP may not have optimal solutions. This is also true for NLPP. A typical result ensuring the existence of optimal solutions is based on the continuity of the functions involved in such a problem.

Theorem 3.2   *Assume that f, g, and h are continuous functions and one of the two following situations holds:*
1. *the set of feasible vectors $g(x) \leq 0$, $h(x) = 0$ is a bounded set in $\mathbf{R}^n$;*
2. *the set of feasible vectors is not bounded, but*

$$\lim_{|x|\to\infty, g(x)\leq 0, h(x)=0} f(x) = +\infty.$$

Then the associated minimization problem admits at least one solution.

There is a further situation in which an NLPP admits an optimal solution, but to elucidate that is part of the aim of this chapter (Section 5).

In many practical settings the above theorem is enough to ensure the existence of an optimal solution. The main topic of this chapter is how to find them.

To better understand how we can find or approximate optimal solutions, we will proceed in two main steps. First, we will treat the case in which all constraints come in the form of equalities:

$$\text{Minimize}\quad f(x)\quad \text{subject to}\quad h(x) = 0.$$

Then, we will examine the general case by appropriately applying the situation of equality constraints. The main issue we would like to understand is, what is special about optimal solutions of NLPP? what must they satisfy in order to be eligible as an optimal solution for a particular optimization problem? This is the question about necessary conditions of optimality, and it will lead to the Karush–Kuhn–Tucker (KKT) conditions. We will investigate several explicit examples. Next, we will be concerned with situations in which those necessary conditions of optimality are indeed sufficient to detect (global) optimal solutions of a NLPP. This will open the whole problem of understanding convexity, and why it is so desirable a property, in minimizing a cost functional under a set of constraints. We will finish the chapter with a brief discussion about duality for NLPP.

In the treatment of NLPP it is important to make a distinction between local and global minima.

Definition 3.3   *A vector* $x^{(0)} \in \mathbf{R}^n$ *is a local minimum of* f *subject to* $g(x) \leq 0, h(x) = 0$ *if*

$$g(x^{(0)}) \leq 0, \quad h(x^{(0)}) = 0,$$

and

$$f(x^{(0)}) \leq f(x)$$

for all x *such that*

$$g(x) \leq 0, \quad h(x) = 0, \quad \left| x - x^{(0)} \right| < \epsilon,$$

for some $\epsilon > 0$.
    A vector $x^{(0)}$ *is a global minimum of* f *subject to* $g(x) \leq 0, h(x) = 0$ *if*

$$g(x^{(0)}) \leq 0, \quad h(x^{(0)}) = 0,$$

and

$$f(x^{(0)}) \leq f(x)$$

for all x *such that*

$$g(x) \leq 0, \quad h(x) = 0.$$

Notice the difference between these two concepts.

2. LAGRANGE MULTIPLIERS

In this section we will try to derive the conditions that a vector must satisfy so that it can possibly be an optimal solution for the problem

$$\text{Minimize} \quad f(x) \quad \text{under} \quad h(x) = 0.$$

It may be possible that some of our readers may already know how to write down the optimality conditions for the above problem, i.e., those equations in terms of f and h that optimal solutions must satisfy. This is sometimes taught in advanced calculus courses. One needs to introduce Lagrange multipliers, which are parameters associated with the constraints, one for each individual constraint. If λ is such a vector of multipliers, then optimal solutions of the NLPP must be solutions for the system of equations

$$\nabla f(x) + \lambda \, \nabla h(x) = 0, \quad h(x) = 0, \tag{3–1}$$

where the pairs (x, λ) of points and multipliers are the unknowns. Notice that we have as many equations as unknowns, $n + m$ altogether if $x \in \mathbf{R}^n$, $\lambda \in \mathbf{R}^m$, and we have m constraints, so that $h : \mathbf{R}^n \to \mathbf{R}^m$. It is important to stress that not all solutions of (3–1) will be optimal solutions for our problem. What is true is that optimal solutions are to be found among the solutions of (3–1). Other solutions of this system may correspond to maxima, local minima and maxima, saddle points, etc.

Theorem 3.4   *Every optimal solution of*

$$\text{Minimize} \quad f(x) \quad \text{under} \quad h(x) = 0$$

must be a solution of the system of necessary conditions of optimality

$$\nabla f(x) + \lambda \, \nabla h(x) = 0, \quad h(x) = 0.$$

Before providing some justification for the conditions of optimality for those interested readers, we are going to look at several examples and see how they can be used to find optimal solutions.

Example 3.5   *We would like to find the extreme values (maximum and minimum) of the function*

$$f(x_1, x_2, x_3) = x_1^3 + x_2^3 + x_3^3$$

$$L(X_1, X_2, X_3, \lambda) = X_1^3 + X_2^3 + X_3^3 + \lambda(4 - X_1^2 - X_2^2 - X$$
$$= X_1^3 + X_2^3 + X_3^3 + 24 - \lambda X_1^2 - \lambda X_2^2 - \lambda X_3$$

over the sphere $x_1^2 + x_2^2 + x_3^2 = 4$. In this case $n = 3$, $m = 1$, and

$$h(x_1, x_2, x_3) = x_1^2 + x_2^2 + x_3^2 - 4.$$

The optimality conditions (3–1) can be written

$$\frac{\partial L}{\partial x_1} = 3x_1^2 + \lambda 2x_1 = 0,$$
$$\frac{\partial L}{\partial x_2} = 3x_2^2 + \lambda 2x_2 = 0,$$
$$3x_3^2 + \lambda 2x_3 = 0,$$
$$\frac{\partial L}{\partial x_3} = x_1^2 + x_2^2 + x_3^2 = 4. = \frac{\partial L}{\partial \lambda}$$

[handwritten: $3x_1^{\,x} = -\lambda 2x_1$ *,* $x_1 = -\frac{\lambda 2}{3}$ *]*

Writing down these equations should not pose any particular difficulty. Finding their solutions may require, however, some computational maturity. Since we have to deal with a nonlinear system of equations, there is no way we can know in advance how many vectors we are looking for, so that in manipulating equations we have to ensure that no solution is lost, since in particular, the optimal solution we are seeking might be precisely the one not found. In our particular situation, factoring out the first three equations, we obtain

$$(3x_1 + \lambda 2)x_1 = 0,$$
$$(3x_2 + \lambda 2)x_2 = 0,$$
$$(3x_3 + \lambda 2)x_3 = 0,$$
$$x_1^2 + x_2^2 + x_3^2 = 4.$$

[handwritten left: (2) $x_3^2 = 4$ *,* $x_3 = \pm 2$ *]*

[handwritten right: (4) $3x_1^2 = 4$ *,* $x_1 = \pm 2/\sqrt{3}$ *]*

The first three equations have a product structure, and so we will have eight possibilities depending on the factors that vanish. Moreover, due to the symmetry of the equations with respect to the three independent variables, it suffices to consider four cases:

1. $x_1 = x_2 = x_3 = 0$: this situation is inconsistent with the constraint;
2. $x_1 = x_2 = 0$, $x_3 \neq 0$: bearing in mind the constraint, we obtain $x_3 = \pm 2$, and the third equation may be used to determine the value of the multiplier (for the moment we are not especially interested in that);
3. $x_1 = 0$, $x_2, x_3 \neq 0$: from the second and third equations we conclude that $x_2 = x_3$, and taking this information into the constraint, $x_2 = x_3 = \pm\sqrt{2}$;
4. $x_1, x_2, x_3 \neq 0$: the first three equations lead to $x_1 = x_2 = x_3$, and the constraint ensures that this common value is $\pm 2/\sqrt{3}$.

In summary, and having in mind the symmetry, we have the following candidates for the maximum and minimum points

$$(\pm 2, 0, 0), \quad (0, \pm 2, 0), \quad (0, 0, \pm 2),$$
$$(0, \sqrt{2}, \sqrt{2}), \quad (0, -\sqrt{2}, -\sqrt{2}), \quad (\sqrt{2}, 0, \sqrt{2}), \quad (-\sqrt{2}, 0, -\sqrt{2}),$$
$$(\sqrt{2}, \sqrt{2}, 0), (-\sqrt{2}, -\sqrt{2}, 0), \quad (2/\sqrt{3}, 2/\sqrt{3}, 2/\sqrt{3}), \quad (-2/\sqrt{3}, -2/\sqrt{3}, -2/\sqrt{3}).$$

On the other hand, the important observation that the sphere is a bounded surface in space enables us to know that indeed the continuous function f ought to attain its two extreme values somewhere. Therefore, the points where the maximum and minimum are assumed must be contained in the preceding list. By simply computing f at those points and comparing their values, we find that the maximum is 8 and is attained at $(2, 0, 0)$, $(0, 2, 0)$, and $(0, 0, 2)$ while the minimum is -8 and corresponds to $(-2, 0, 0)$, $(0, -2, 0)$, and $(0, 0, -2)$.

Example 3.6 Let us now assume that we might be interested in knowing the extreme values (maximum and minimum) of the same function f not over all the points of the sphere but only over those that lie at the same time in the plane $x_1 + x_2 + x_3 = 1$, that is, we would like to find the extreme points of the function

$$f(x_1, x_2, x_3) = x_1^3 + x_2^3 + x_3^3$$

over the set of points satisfying

$$x_1^2 + x_2^2 + x_3^2 = 4, \quad x_1 + x_2 + x_3 = 1.$$

The equations of optimality this time are

$$3x_1^2 + \lambda_1 2x_1 + \lambda_2 = 0,$$
$$3x_2^2 + \lambda_1 2x_2 + \lambda_2 = 0,$$
$$3x_3^2 + \lambda_1 2x_3 + \lambda_2 = 0,$$
$$x_1^2 + x_2^2 + x_3^2 = 4,$$
$$x_1 + x_2 + x_3 = 1,$$

a system of five equations and five unknowns $x_1, x_2, x_3, \lambda_1, \lambda_2$. Since we are not particularly interested in finding the values of the multipliers, and the first three equations imply that the gradient of f must be a linear combination of the gradients of the two restrictions, we can rewrite the previous system as

$$\begin{vmatrix} 3x_1^2 & 2x_1 & 1 \\ 3x_2^2 & 2x_2 & 1 \\ 3x_3^2 & 2x_3 & 1 \end{vmatrix} = 0,$$
$$x_1^2 + x_2^2 + x_3^2 = 4,$$
$$x_1 + x_2 + x_3 = 1.$$

After factoring out 3 in the first column and 2 in the second, we arrive at a Vandermonde determinant whose expression is well known

$$(x_1 - x_2)(x_1 - x_3)(x_2 - x_3) = 0,$$
$$x_1^2 + x_2^2 + x_3^2 = 4,$$
$$x_1 + x_2 + x_3 = 1.$$

It is elementary to find the solutions of this system. There are three distinct possibilities, which due to symmetry, reduce to essentially one:

$$x_1 = x_2, \quad x_3 = 1 - 2x_1, \quad 2x_1^2 + (1 - 2x_1)^2 = 4.$$

The other solutions are obtained by permutations of the variables. The explicit solutions are

$$\left(\frac{1}{3} + \frac{\sqrt{22}}{6}, \frac{1}{3} + \frac{\sqrt{22}}{6}, \frac{1}{3} - \frac{\sqrt{22}}{3} \right),$$

$$\left(\frac{1}{3} - \frac{\sqrt{22}}{6}, \frac{1}{3} - \frac{\sqrt{22}}{6}, \frac{1}{3} + \frac{\sqrt{22}}{3} \right),$$

and those obtained by permutations of these. It is straightforward to check that the first set of solutions correspond to the maximum value, and the others to the minimum. Notice how these solutions differ from those of Example 3.5.

How can we understand where multipliers come from? How do they arise? A simple way of understanding this is by considering parametrized curves

$$\tau : (-\delta, \delta) \to \mathbf{R}^n, \quad \delta > 0,$$

whose image $\tau(-\delta, \delta)$ is entirely contained in the feasible set of our optimization problem; that is

$$h(\tau(t)) = 0, \quad \text{for all } t \in (-\delta, \delta).$$

If we suppose that $x_0 \in \mathbf{R}^n$ is a point of local minimum or maximum, or even a saddle point with respect to vectors in the feasible set, and assume that τ passes through x_0 for $t = 0$, $\tau(0) = x_0$, then the composition $f(\tau(t))$ must likewise have a local minimum, local maximum, or saddle point for $t = 0$. What characterizes any of these situations is that the derivative must vanish. By the chain rule,

$$0 = \left.\frac{df(\tau(t))}{dt}\right|_{t=0} = \nabla f(\tau(0)) \, \tau'(0) = \nabla f(x_0) \, \tau'(0).$$

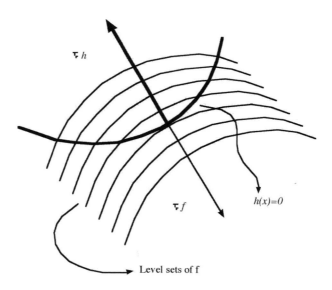

Figure 3.1. Lagrange multipliers.

On the other hand, since $h(\tau(t)) = 0$ for all t, we should also have in the same way

$$0 = \nabla h(x_0) \, \tau'(0).$$

Since the tangent vector $\tau'(0)$ is arbitrary, as far as it respects these two equalities, we conclude that these can hold simultaneously if and only if $\nabla f(x_0)$ belongs to the span of $\nabla h(x_0)$. This linear dependence gives rise to the multipliers. See Figure 3.1.

It is important to point out that the method of multipliers may fail in providing the extreme points when some of the constraints $h_i(x) = 0$ represent a surface (or hypersurface) that is not regular in the sense that its gradient vector ∇h_i vanishes at some point, or when the intersection of the sets $h_i(x) = 0$ is somehow not regular. These points are thus called singular, and we should include them in the list of candidates for maximum and/or minimum points. This issue (nonsmooth optimization) is, however, beyond the scope of this text. See [8].

Example 3.7 *We would like to determine the minimum value that the expression*

$$y = \sum_{i=1}^{n} a_i x_i^2$$

can attain with respect to the variables x_i ($a_i > 0$ are given numbers) under the constraint

$$c = \sum_{i=1}^{n} x_i,$$

where c is another given constant. Optimality conditions lead to

$$2a_j x_j + \lambda = 0, \quad j = 1, 2, \ldots, n,$$

so that

$$x_j = -\frac{\lambda}{2a_j}.$$

If we take these expressions back into the constraint

$$c = -\sum_{i=1}^{n} \frac{\lambda}{2a_i},$$

then

$$\lambda = -\frac{2c}{\displaystyle\sum_{i=1}^{n} \frac{1}{a_i}},$$

and consequently,

$$x_j = \frac{c}{a_j \displaystyle\sum_{i=1}^{n} \frac{1}{a_i}},$$

which is the only solution of the system. Notice that since the objective function tends to $+\infty$ when some of the variables grow indefinitely, the attainment of the minimum value is guaranteed. Therefore, the solution found must correspond to the minimum. The maximum is $+\infty$, since the constraint is not able to keep all variables from growing indefinitely.

A more sophisticated, but instructive, example follows.

Example 3.8   Let a be a fixed vector in $\mathbf{R}^3$. We want to determine the extreme values of the linear cost function ax under the constraints

$$x_1 x_2 + x_1 x_3 + x_2 x_3 = 0,$$
$$|x|^2 = x_1^2 + x_2^2 + x_3^2 = 1.$$

For simplicity, we will use the notation

$$\det x = x_1 x_2 + x_1 x_3 + x_2 x_3.$$

Notice that the set of vectors x satisfying the two constraints is a subset of the unit sphere in $\mathbf{R}^3$, so that it is bounded, and the cost functional necessarily attains its maximum and minimum values. These can be detected by examining the necessary conditions of optimality, namely,

$$a + \lambda_1 A x + \lambda_2 x = 0,$$
$$\det x = 0,$$
$$|x|^2 = 1.$$

The matrix A is

$$A = \begin{pmatrix} 0 & 1 & 1 \\ 1 & 0 & 1 \\ 1 & 1 & 0 \end{pmatrix}.$$

The first (vector) equation informs us that the vector a must be a linear combination (whose coefficients are the multipliers) of the other two vectors, Ax and *x. By eliminating the multipliers, we could equivalently write this equation by requiring*

$$0 = \begin{vmatrix} a \\ Ax \\ x \end{vmatrix}.$$

But since

$$0 = \begin{vmatrix} a \\ x \\ x \end{vmatrix},$$

we can also have

$$0 = \begin{vmatrix} a \\ Ax \\ x \end{vmatrix} + \begin{vmatrix} a \\ x \\ x \end{vmatrix} = \begin{vmatrix} a \\ x + Ax \\ x \end{vmatrix}.$$

Let $e = (1, 1, 1)$. Notice that $x + Ax = (x\ e)e$, so that

$$0 = xe \begin{vmatrix} a \\ e \\ x \end{vmatrix}.$$

Observe that xe can never vanish because for one of our feasible vectors x we have

$$|xe|^2 = |x|^2 + 2\det x = 1.$$

Therefore, we must have

$$\begin{vmatrix} a \\ e \\ x \end{vmatrix} = 0,$$

and this implies

$$x = sa + te$$

for certain coefficients s and t. If we take this expression into the two constraints $|x|^2 = 1$ and $\det x = 0$, after a few computations we get the two quadratic equations

$$\det a\ s^2 + 2st\ a\ e + 3t^2 = 0,$$

$$|a|^2 s^2 + 2st\ a\ e + 3t^2 = 1.$$

We immediately obtain

$$s^2 \left(|a|^2 - \det a \right) = 1.$$

Note that (why?)

$$|a|^2 - \det a \geq 0.$$

If

$$|a|^2 - \det a = 0,$$

then the equation for s is inconsistent. In fact, in this situation we do not have any solution for the optimality equations. However, this situation can occur only when a is a multiple of e (why?), so that the cost function is

$$cxe,$$

and as indicated earlier,

$$|xe|^2 = |x|^2 + 2\det x = 1,$$

and therefore the cost functional is constant for all feasible vectors.

Assume that a is such that

$$|a|^2 - \det a > 0.$$

In this case we have two solutions for s:

$$s = \frac{\pm 1}{\sqrt{|a|^2 - \det a}}.$$

Taking these values into the first of the quadratic equations above and solving for t leads to (using again the formula for $(ae)^2$)

$$t = \pm\frac{1}{3} \pm \frac{ae}{3\sqrt{|a|^2 - \det a}}.$$

The valid pairs of solutions are

$$\left(\frac{1}{\sqrt{|a|^2 - \det a}}, \pm\frac{1}{3} - \frac{a\,e}{3\sqrt{|a|^2 - \det a}} \right), \quad \left(\frac{-1}{\sqrt{|a|^2 - \det a}}, \pm\frac{1}{3} + \frac{a\,e}{3\sqrt{|a|^2 - \det a}} \right)$$

Among these four vectors we have to find the maximum and minimum values of the inner product

$$ax = s\,|a|^2 + tae.$$

By examining carefully the four possibilities, we conclude that the maximum value is

$$\frac{1}{3}\left(2\sqrt{|a|^2 - \det a} + |ae|\right),$$

and it is attained at

$$x = sa + te$$

for

$$s = \frac{1}{\sqrt{|a|^2 - \det a}}, \qquad t = \frac{ae}{|ae|}\frac{1}{3} - \frac{ae}{3\sqrt{|a|^2 - \det a}}.$$

The minimum value is the maximum changed in sign, and it is attained at the opposite of the point of maximum.

3. KARUSH–KUHN–TUCKER OPTIMALITY CONDITIONS

We would like to treat the general case in which some of the constraints come in the form of equalities and some come in the form of inequalities:

Minimize $f(x)$ subject to $g(x) \le 0$, $h(x) = 0$.

Let us first explore what sort of conditions a point needs to satisfy so that it can be an optimal solution of our problem. What is special about such a point?

Let $x^{(0)}$ be one such point of minimum, and let M be the set of indices $M = \{1, 2, \ldots, m\}$, where m is precisely the number of components of g. We consider the following subset of M:

$$J = \left\{ j \in M : g_j(x^{(0)}) = 0 \right\}.$$

It might well happen that this set is empty. For j belonging to $M \setminus J$, we say that the corresponding constraint is nonbinding or inactive. Let us look at the auxiliary problem

Minimize $f(x)$ under $g_j(x) = 0$, $j \in J$, $h(x) = 0$.

Our initial solution $x^{(0)}$ will certainly be a point of local minimum, perhaps not global (why?), and consequently, since all constraints for this new problem are in the form of equalities, there exists a collection of multipliers

$$\mu_j, \quad j \in J, \quad \lambda \in \mathbf{R}^d,$$

such that

$$\nabla f(x^{(0)}) + \sum_{j \in J} \mu_j \nabla g_j(x^{(0)}) + \lambda \nabla h(x^{(0)}) = 0. \tag{3–2}$$

Furthermore, we assert that μ_j can be taken to be nonnegative. The intuitive reason for this is that f has a minimum at the point $x^{(0)}$ but each of the g_j has a maximum, because $g_j(x^{(0)}) = 0$ is the maximum value that g_j can attain in the feasible set for our initial problem. Hence the gradients of f and g_j at the point $x^{(0)}$ "must point in different directions." This assertion actually requires more rigor and care, but it is enough for our purposes. For $j \in M \setminus J$, we take $\mu_j = 0$. We thus arrive at the necessary conditions of optimality, known as Karush–Kuhn–Tucker (KKT) conditions.

Theorem 3.10   *If x is a nonsingular optimal solution of our problem, then there exists a vector of multipliers (μ, λ) such that*

$$\nabla f(x) + \mu \nabla g(x) + \lambda \nabla h(x) = 0,$$
$$\mu g(x) = 0,$$
$$\mu \geq 0, \quad g(x) \leq 0, \quad h(x) = 0.$$

The necessity of such conditions means that optimal solutions must be sought among those vectors x for which we can find a set of multipliers (μ, λ) satisfying the preceding conditions. This information lets us select those points that are feasible for minimum points. In those situations in which all such solutions may be found, and we have the information that our problem actually must have at least one solution, these can be identified by simply computing the cost of all such candidates and deciding on the minimum.

Before analyzing several explicit examples, we would like to make a couple of interesting observations.

1. If the optimization problem consists in finding the maximum instead of the minimum, playing with the appropriate minus signs, it is not difficult to

write down the changes in the KKT conditions. In fact, we should have

$$\nabla f(x) + \mu \nabla g(x) + \lambda \nabla h(x) = 0,$$
$$\mu g(x) = 0,$$
$$\mu \leq 0, \quad g(x) \leq 0, \quad h(x) = 0.$$

If we do not care about the sign of the components of μ, the list of feasible points for extrema will significantly increase with solutions that cannot be either maxima or minima, since the signs of components of μ will be mixed up positive and negative. Therefore, if all components of μ are nonnegative, the corresponding point may possibly be a point of minimum (never a maximum); if all components are nonpositive, the point may possibly be a point of maximum (never a minimum); and if there are positive and negative components of μ, then the point can never be either a point of maximum or of minimum (saddle point).

2. The conditions

$$\mu \geq 0, \quad g(x) \leq 0, \quad \mu g(x) = 0,$$

are equivalent (why?) to

$$\mu \geq 0, \quad g(x) \leq 0, \quad \mu_j g_j(x) = 0, \ j = 1, 2, \ldots, m,$$

so that to find all solutions for the KKT conditions, we have to look for all solutions of the system of $n + m + d$ equations in $n + m + d$ unknowns x, μ, λ,

$$\nabla f(x) + \mu \nabla g(x) + \lambda \nabla h(x) = 0,$$
$$\mu_j g_j(x) = 0, \quad j = 1, 2, \ldots, m,$$
$$h_i(x) = 0, \quad i = 1, 2, \ldots, d,$$

that also satisfy $\mu \geq 0$, $g(x) \leq 0$, in the case we are interested in the minimum; and that satisfy $\mu \leq 0$, $g(x) \leq 0$, in the case of the maximum. When trying to solve the previous system, and due to its particular structure, we can always proceed by examining the 2^m cases

$$\mu_i = 0, \quad g_j(x) = 0, \quad j \in J, i \in M \setminus J,$$

where J runs through all possible subsets of $M = \{1, 2, \ldots, m\}$. Among all different possibilities that we may obtain, we must discard those that are impossible or those that we are not interested in. Although there can be other reasonable methods of solving the system, this is a rational way of organizing computations. On the other hand, any observation based on the particular nature of the problem, leading to discarding some of the solutions, may also simplify considerably the solution-finding procedure.

Let us look at several examples.

Example 3.10 *Suppose that a certain electrical network consists of three different channels through which electric power flows. If x_i, $i = 1, 2, 3$, stands for the amount of power through channel i, the total loss in the network is given by the function*

$$p(x_1, x_2, x_3) = x_3 + \frac{1}{2}\left(x_1^2 + x_2^2 + \frac{x_3^2}{10}\right).$$

If a total amount of r is to be transferred, determine the amounts through each channel so as to minimize the loss of power. Evidently, the problem reduces to finding the minimum of the above function providing a measure of the loss of power under the constraints

$$x_1 + x_2 + x_3 = r, \quad x_i \geq 0.$$

Since the set of points of $\mathbf{R}^3$ satisfying these constraints is bounded (why?), the point of minimum we are looking for should be one of the solutions of the system of optimality

$$x_1 - \mu_1 + \lambda = 0, \quad x_2 - \mu_2 + \lambda = 0, \quad 1 + \frac{x_3}{10} - \mu_3 + \lambda = 0,$$
$$\mu_1 x_1 = 0, \quad \mu_2 x_2 = 0, \quad \mu_3 x_3 = 0, \quad x_1 + x_2 + x_3 = r,$$

where the unknowns are $x_1, x_2, x_3, \mu_1, \mu_2, \mu_3, \lambda$. We are interested in those solutions satisfying $x_1, x_2, x_3, \mu_1, \mu_2, \mu_3 \geq 0$. Notice that λ is associated with the equality constraint $x_1 + x_2 + x_3 = r$, and thus we cannot demand any condition on its sign. Solving the previous system requires a little bit of ability in finding all the solutions corresponding to the eight possibilities

$$x_1 = x_2 = x_3 = 0,$$
$$x_1 = x_2 = \mu_3 = 0,$$
$$x_1 = \mu_2 = x_3 = 0,$$
$$\mu_1 = x_2 = x_3 = 0,$$
$$x_1 = \mu_2 = \mu_3 = 0,$$
$$\mu_1 = x_2 = \mu_3 = 0,$$
$$\mu_1 = \mu_2 = x_3 = 0,$$
$$\mu_1 = \mu_2 = \mu_3 = 0.$$

After studying with a little bit of care all these possibilities and discarding those solutions we are not interested in, we arrive at the optimal solution

$$(r/2, r/2, 0) \quad \text{with associated multipliers} \quad (0, 0, 1 - r/2, -r/2)$$

when $r \leq 2$, and

$$((10 + r)/12, (10 + r)/12, 5(r - 2)/6) \quad \text{with multipliers} \quad (0, 0, 0, -(10 + r)/12)$$

for $r \geq 2$. Notice how both solutions coincide when $r = 2$.

Example 3.11 We have a rope of length a to tie a box from top to bottom along the two perpendicular directions. What is the maximum volume that such a box can contain?

We would like to determine the maximum of the volume function

$$V(x_1, x_2, x_3) = x_1 x_2 x_3$$

subject to the conditions

$$x_1, x_2, x_3 \geq 0, \quad 2x_1 + 2x_2 + 4x_3 \leq a,$$

assuming that x_3 is the height. With some care about the minus signs that must be introduced to transform the problem to our standard format, we have the system

$$x_2 x_3 - \mu_1 + 2\mu_4 = 0, \quad x_1 x_3 - \mu_2 + 2\mu_4 = 0, \quad x_1 x_2 - \mu_3 + 4\mu_4 = 0,$$
$$x_1 \mu_1 = 0, \quad x_2 \mu_2 = 0, \quad x_3 \mu_3 = 0, \quad (2x_1 + 2x_2 + 4x_3 - a)\mu_4 = 0.$$

We look for those solutions with

$$x_1, x_2, x_3 \geq 0,$$
$$2x_1 + 2x_2 + 4x_3 \leq a,$$
$$\mu_1, \mu_2, \mu_3, \mu_4 \leq 0.$$

Since the first three equations enable us to express μ_1, μ_2, and μ_3 in terms of x_1, x_2, x_3, and μ_4, we can eliminate the first variables and obtain an equivalent system

$$x_1(x_2 x_3 + 2\mu_4) = 0,$$
$$x_2(x_1 x_3 + 2\mu_4) = 0,$$
$$x_3(x_1 x_2 + 4\mu_4) = 0,$$
$$(2x_1 + 2x_2 + 4x_3 - a)\mu_4 = 0.$$

If we add the first three equations and bear in mind the last one, we arrive at

$$0 = 3x_1 x_2 x_3 + a\mu_4,$$

whence

$$\mu_4 = -3x_1 x_2 x_3/a.$$

If we apply this identity to the same three last equations, and notice that the maximum of V cannot vanish ($x_1 x_2 x_3 \neq 0$), since this would rather be the minimum, we obtain the unique solution for the maximum

$$x_1 = x_2 = a/6, \quad x_3 = a/12.$$

Moreover, the associated multipliers are

$$(0, 0, 0, -a^2/144),$$

which indeed correspond to a point of maximum. Since the constraint region is bounded, this is the optimal solution sought.

Example 3.12   We would like to find the minimum and maximum of

$$f(x_1, x_2, x_3) = x_1^3 + x_2^3 + x_3^3$$

over the region determined by the constraints

$$x_1^2 + x_2^2 + x_3^2 \leq 4, \quad x_1 + x_2 + x_3 \leq 1.$$

It is a simple exercise to write down the KKT conditions for this situation, namely,

$$3x_1^2 + \mu_1 2x_1 + \mu_2 = 0,$$
$$3x_2^2 + \mu_1 2x_2 + \mu_2 = 0,$$
$$3x_3^2 + \mu_1 2x_3 + \mu_2 = 0,$$
$$\mu_1(x_1^2 + x_2^2 + x_3^2 - 4) = 0,$$
$$\mu_2(x_1 + x_2 + x_3 - 1) = 0,$$
$$x_1^2 + x_2^2 + x_3^2 \leq 4,$$
$$x_1 + x_2 + x_3 \leq 1.$$

In addition, we must keep in mind the constraints on the signs of the multipliers, μ_1 and μ_2, when looking for the maximum or the minimum: $\mu_1, \mu_2 \geq 0$ for the minimum, and $\mu_1, \mu_2 \leq 0$ for the maximum. We organize the discussion of the previous system in four cases.

1. $\mu_1 = \mu_2 = 0$: in this case we immediately obtain the solution $x_1 = x_2 = x_3 = 0$, which is admissible for both the maximum and the minimum;
2. $\mu_1 = 0$, $x_1 + x_2 + x_3 = 1$: it is straightforward to get

$$\mu_2 = -3x_1^2 = -3x_2^2 = -3x_3^2,$$

whence $(1/3, 1/3, 1/3)$ is the unique solution, which is admissible for the maximum, not for the minimum, since $\mu_2 = -1/3$;
3. $\mu_2 = 0$, $x_1^2 + x_2^2 + x_3^2 = 4$: adding the first three equations (after eliminating μ_2) and keeping in mind that the sum of squares is unity, we obtain

$$12 + 2(x_1 + x_2 + x_3)\mu_1 = 0.$$

This identity discards the possibility that $x_1 + x_2 + x_3 = 0$, and so

$$\mu_1 = \frac{-6}{x_1 + x_2 + x_3}.$$

If we apply this equality to the first three equations of the optimality system, we arrive at the fact that either the coordinates of the points vanish or else their common value is

$$\frac{4}{x_1 + x_2 + x_3}.$$

From here, the following solutions arise (discarding simultaneously those solutions not satisfying $x_1 + x_2 + x_3 \le 1$):

$$(-2, 0, 0), \quad (0, -2, 0), \quad (0, 0, -2),$$

$$\left(-\sqrt{2}, -\sqrt{2}, 0\right) \quad \left(-\sqrt{2}, 0, -\sqrt{2}\right) \quad \left(0, -\sqrt{2}, -\sqrt{2}\right),$$

$$\left(\frac{-2}{\sqrt{3}}, \frac{-2}{\sqrt{3}}, \frac{-2}{\sqrt{3}}\right).$$

Since all these solutions satisfy $x_1 + x_2 + x_3 < 0$ ($\mu_1 > 0$), they will be feasible for the minimum.

4. The last case is associated with the equalities

$$x_1 + x_2 + x_3 = 1, \quad x_1^2 + x_2^2 + x_3^2 = 4,$$

which has been solved in the previous section.

After computing the values of the cost function f in all those selected points, we come to the conclusion that the maximum is taken at

$$\left(\frac{1}{3} + \frac{\sqrt{22}}{6}, \frac{1}{3} + \frac{\sqrt{22}}{6}, \frac{1}{3} - \frac{\sqrt{22}}{3}\right),$$

and the minimum value is attained at

$$(-2, 0, 0), \quad (0, -2, 0), \quad (0, 0, -2).$$

Compare these results with those of Examples 3.5 and 3.6.

4.   CONVEXITY

We have developed a basic understanding of how to detect necessary conditions of optimality that must be satisfied at a point of (local) maximum or minimum.

We have also sufficiently stressed the fact that solutions to the KKT conditions may include other points that are not the solutions we are looking for. There might exist solutions to the KKT optimality conditions that do not correspond to the extreme values. The fundamental question we would like to address is whether there is some further requirement on the objective function and/or the functions expressing the constraints so that we can ensure that solutions of the KKT conditions are exactly the points where the minimum (or maximum) is attained, without a discussion "a posteriori" on the nature of the different solutions. As we will understand later, this is a vital issue since in most of the situations one encounters in practice, solutions for the KKT conditions cannot be explicitly found and need to be approximated. One is never sure whether all solutions have been found. Because of the relevance of this issue, we will analyze the situation in greater detail starting with the most basic case in nonlinear programming so that we may grasp the whole point of the notion of convexity.

Let us consider the problem

$$\text{Minimize} \quad f(x), \qquad x \in \mathbf{R},$$

assuming that f is as regular as we may need. The KKT condition reduces in this simplified situation to

$$f'(x) = 0.$$

This (nonlinear) equation might have many solutions, even infinitely many, and any one of them could be the point of minimum sought. But some of those could also correspond to points of local minima, local maxima, saddle points, etc. Let us imagine the following situation: We know that

$$f(x) \to +\infty \text{ when } x \to \pm\infty,$$
$$f'(x) = 0 \quad \text{has a unique solution} \quad x_0.$$

It is not difficult to realize that x_0 is truly the point of global minimum, and therefore the unique solution of our initial minimization problem (why?). The first requirement on the infinite limits of f can be more or less easy to check once we know what f is. But how can we be sure about the uniqueness of the solution of the equation of critical points without having to solve it? If we assume that f admits a continuous second derivative $f''(x)$, and $f''(x) > 0$ for all x (which in many instances can be easily checked), then we can ensure the second requirement: The equation $f'(x) = 0$ can have only one solution (why?).

On the other hand, we know from elementary calculus that the condition $f'' > 0$ means that f is convex. In summary:

1. $f(x) \to +\infty$, $x \to \pm\infty$: the equation $f'(x) = 0$ has at least one solution corresponding to the global minimum of f;
2. $f''(x) > 0$ for all x: f is strictly convex, and the equation $f'(x) = 0$ has at most one solution.

Consequently, if both requirements hold, the only solution of the equation $f'(x) = 0$ will correspond to the global minimum, and in particular, there is no local minimum that is not global. This is basically the reason why convexity is so desperately needed in treating minimization problems (the same is true with concavity for maximum problems). Another way of summing up the preceding remarks is by saying that under convexity, necessary conditions of optimality become sufficient, because convexity rules out the existence of local minima that are not global. Since we are in front of one of the main concepts in optimization, we are going to treat it with a little bit of care.

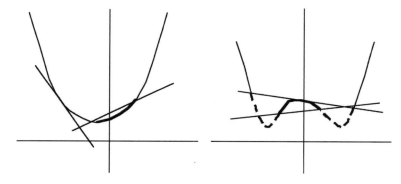

Figure 3.2. A convex and a nonconvex function.

Definition 3.13   *A set K in $\mathbf{R}^n$ is convex if for every pair of vectors $x, y \in K$, the segment joining them is also contained in K:*

$$tx + (1-t)y \in K, \quad t \in [0,1].$$

A function $f : K \subset \mathbf{R}^n \to \mathbf{R}$ is said to be convex if K is a convex set of vectors and

$$f(tx + (1-t)y) \le tf(x) + (1-t)f(y), \quad \text{whenever } x, y \in K, \quad t \in [0,1].$$

Before proceeding to attempting a better understanding of this condition, it is worthwhile to be persuaded that it is a key concept relevant for minimization problems. We will learn more about convexity in the next section. Our readers will most likely know what the convexity condition means geometrically. See Figure 3.2.

The reason why convexity is so important in minimization problems can be formulated as follows.

Theorem 3.14   *Let*
$$f : K \subset \mathbf{R}^n \to \mathbf{R}$$

be convex where K is also convex. If x_0 is a local minimum for f in K, then it is also a global minimum for f in K.

For a justification of this result, notice that the condition of x_0 being a local minimum for f in K means that

$$f(x_0) \le f(x), \quad |x - x_0| < \epsilon, x \in K,$$

where $\epsilon > 0$ is some given number. Let us put

$$B_\epsilon = \{x \in K : |x - x_0| < \epsilon\}.$$

Let $y \in K$ be an arbitrary point in K. The segment joining x_0 and y undoubtedly has points belonging to B_ϵ,

$$tx_0 + (1 - t)y \in B_\epsilon, \quad t \text{ sufficiently close to 1.}$$

Hence for such t's and by the convexity of f,

$$f(x_0) \le f(tx_0 + (1 - t)y) \le tf(x_0) + (1 - t)f(y).$$

Rearranging these terms, we have

$$(1 - t)f(x_0) \le (1 - t)f(y).$$

Since $1 - t > 0$ for some such t, we conclude that

$$f(x_0) \le f(y).$$

The arbitrariness of $y \in K$ yields the desired result.

The definition of convexity (Definition 3.13) means that for every couple of points in K, x, y, the values of f along the points in the segment joining them do not exceed those of the "line" through $(x, g(x)), (y, g(y))$; said differently, the values of f are under each one of its (f's) secants. If the function f is differentiable, then an alternative characterization of convexity can be given. This is the most appropriate in our context because it can be directly related to optimality conditions. Even further, if the function f is twice differentiable with continuous Hessian matrix, then one can also verify convexity in terms of second derivatives.

Proposition 3.15   *Let*

$$f : K \subset \mathbf{R}^n \to \mathbf{R}$$

be a continuous function where K is convex and open.
1. If f is differentiable and ∇f is continuous, then f is convex if and only if

$$f(y) \geq f(x) + \nabla f(x) \, (y - x), \quad x, y \in K. \tag{3--3}$$

2. If f is twice differentiable and $\nabla^2 f$ is continuous, then f is convex if and only if $\nabla^2 f(x)$ is positive semidefinite for all $x \in K$.

Since the expression
$$f(x) + \nabla f(x) \, (y - x),$$
considered as a function of $y \in K$, is the equation of the tangent hyperplane of f at x, the inequality (3--3) says that the graph of f stays above any of its tangent hyperplanes. Concerning the characterization with the second derivatives, we can equivalently say that if f is twice differentiable and $\nabla^2 f$ is continuous, then it is convex if and only if the eigenvalues of $\nabla^2 f(x)$ are nonnegative at every $x \in K$.

The proof of Proposition 3.15 proceeds in two steps. First, we will try to show that it is true for functions of a single variable $h : J \to \mathbf{R}$, where J is an interval in $\mathbf{R}$.

Let us therefore assume that such an h is differentiable and satisfies

$$h(tx + (1 - t)y) \leq th(x) + (1 - t)h(y), \quad x, y \in J, \quad t \in [0, 1].$$

By rearranging and manipulating terms, we can transform it (when $1 - t > 0$ and $x \neq y$) to

$$(y - x)\frac{h(x + (1 - t)(y - x)) - h(x)}{(1 - t)(y - x)} \leq h(y) - h(x).$$

Since this inequality is correct for every $t \in [0, 1]$, by taking limits as $t \to 1^-$, we can conclude that

$$(y - x)h'(x) \leq h(y) - h(x).$$

This is the first part of the proposition. If we further assume that h has a second derivative at every point of J where the above inequalities hold, we will have, depending on whether $y > x$ or $x > y$,   ·

$$\frac{h(y) - h(x)}{y - x} \geq h'(x), \quad \frac{h(y) - h(x)}{y - x} \leq h'(x).$$

By the mean value theorem applied to h' we ascertain the existence of z such that

$$h'(z) \geq h'(x), \ z \geq x, \quad \text{or} \quad h'(z) \leq h'(x), \ z \leq x.$$

The arbitrariness of y leads to the arbitrariness of z, and hence h' is a nondecreasing function that translates in the nonnegativity of h''. This is the criterion for convexity when functions are twice differentiable.

We finally argue that if h is a twice differentiable function, and its second derivative is nonnegative, then we must necessarily have

$$h(tx + (1 - t)y) \leq th(x) + (1 - t)h(y), \quad x, y \in J, \quad t \in [0, 1].$$

To this aim we rearrange terms in the expression

$$th(x) + (1 - t)h(y) - h(tx + (1 - t)y)$$

in the following fashion:

$$t\left(h(x) - h(tx + (1 - t)y)\right) + (1 - t)\left(h(y) - h(tx + (1 - t)y)\right)$$
$$= t(1 - t)(x - y)h'(a) - t(1 - t)(x - y)h'(b),$$ ·

where we have used the mean value theorem, and points a and b lie between x and $tx + (1 - t)y$, and $tx + (1 - t)y$ and y, respectively. Again by the mean value theorem applied to h' we show the existence of a number c between x and y such that

$$th(x) + (1 - t)h(y) - h(tx + (1 - t)y) = t(1 - t)(x - y)(a - b)h''(c).$$

If we notice that the product $(x - y)(a - b)$ is always nonnegative (even though the two factors could be negative), the nonnegativity of the remaining factors leads us to have

$$th(x) + (1 - t)h(y) - h(tx + (1 - t)y) \geq 0$$

as desired.

For the second step, consider a function $f : K \to \mathbf{R}$ of several variables. In fact, the proof of this case is based on what we have already shown for functions of one variable by simply applying the previous conclusions to the sections

$$h(s) = f(x + s(y - x))$$

for fixed x, y, and conveniently applying the chain rule to compute derivatives and second derivatives. We leave the details to the interested reader.

Sometimes it is important, because it leads to significant consequences, to know about strict inequalities in all three ways of checking convexity. Functions enjoying this additional requirement are called strictly convex, and we talk about strict convexity.

Definition 3.16   *A function $f : K \to \mathbf{R}$, where $K \subset \mathbf{R}^n$ is convex, is called strictly convex if f is continuous (this is in fact redundant) and*

$$f(tx + (1 - t)y) < tf(x) + (1 - t)f(y), \quad x \neq y \in K, \quad t \in (0, 1).$$

A characterization similar to Proposition 3.15 can be shown for strict convexity when the function f is more regular.

Proposition 3.17   *Let*

$$f : K \subset \mathbf{R}^n \to \mathbf{R}$$

be a continuous function where K is convex and open.

1. If f is differentiable and ∇f is continuous, then f is strictly convex if and only if

$$f(y) > f(x) + \nabla f(x) (y - x), \quad x \neq y \in K;$$

2. If f is twice differentiable and $\nabla^2 f$ is continuous, then f is stricly convex if and only if its Hessian matrix is positive definite at every point in K; alternatively, all eigenvalues are strictly positive at every point in K; or even, by Sylvester's criterion, the principal subdeterminants are strictly positive at every point of K.

The proof is an interesting exercise in going over the proof of Proposition 3.15 and checking inequalities and strict inequalities. As a general rule, we can say that convexity lacking strict convexity is typically associated with "flat parts of the graph."

We end this section by looking at several examples of convex functions.

Example 3.18 *Every linear (or affine) function is convex but not strictly convex.*

There are four elementary operations that respect convexity:

1. a linear combination of convex functions with nonnegative coefficients is again a convex function;
2. if $T : \mathbf{R}^N \to \mathbf{R}^m$ is linear, and $g : \mathbf{R}^m \to \mathbf{R}$ is convex, the composition $f(x) = g(Tx)$ is also convex;
3. if $g : K \subset \mathbf{R}^N \to \mathbf{R}$ is convex and $h : \mathbf{R} \to \mathbf{R}$ is convex and nondecreasing, the composition $f(x) = h(g(x))$ is also convex;
4. the supremum of any family of convex functions is again a convex function.

These four statements are easy to check by using directly the definition of convexity itself.

By using linear functions and the basic operations we have listed above, we can generate new convex functions, or deduce the convexity of known examples. For instance, if we realize that

$$|x| = \sup \{a \ x : |a| = 1\},$$

it turns out that the distance to the origin, $|x|$, is a convex function. Moreover, since

$$h(t) = t^p, \quad t \geq 0,$$

is a convex increasing function if $p \geq 1$,

$$g(x) = |x|^p$$

is convex if $p \geq 1$. This function is strictly convex if $p > 1$. If we take

$$h(t) = \sqrt{1 + t^2},$$

which is again a convex nondecreasing function when $t \geq 0$, the function

$$g(x) = \sqrt{1 + |x|^p}$$

will be convex if $p \geq 2$, and strictly convex if $p > 2$. The functions

$$|x|^p + |x|^q, \quad a\,x + \sqrt{1 + |x|^p},$$
$$|x - a|^p, \quad |a\,x|^p + |b\,x|^q,$$

are convex when the exponents p, q are greater or equal to 1.

If a function f is not convex, we define its convexification by putting

$$Cf(x) = \sup\{h(x) : h \leq g, \ h, \ \text{convex}\},$$

which is the greatest convex function among all those under f. If f is already convex, its convexification is f itself. For instance, if f is given by

$$f(x) = \min\{(x + 1)^2, (x - 1)^2\},$$

which is not convex, its convexification is the function defined piecewise by

$$g(x) = \begin{cases} (x + 1)^2, & x \leq -1, \\ 0, & |x| \leq 1, \\ (x - 1)^2, & x \geq 1, \end{cases}$$

a convex, not strictly convex function.

5. SUFFICIENCY OF THE KKT CONDITIONS

We have learned in the preceding section about the convexity condition, trying to emphasize its relevance concerning minimization problems. In this section, we would like to apply those ideas to the particular situation of NLPP, and in particular show that necessary conditions of optimality become sufficient as well, under the main assumption of convexity of all functions involved. Since optimality conditions are formulated in terms of first derivatives, the appropriate notion of convexity is the one in which first derivatives appear: A function $f : K \subset \mathbf{R}^n \to \mathbf{R}$ is convex if K is a convex set of vectors and

$$f(y) \geq f(x) + \nabla f(x)\,(y - x), \quad y, x \in K, t \in [0, 1].$$

As usual, our model problem is

$$\text{Minimize} \quad f(x) \quad \text{subject to} \quad g(x) \leq 0, h(x) = 0,$$

where we explicitly assume that f, g, and h are defined on all of $\mathbf{R}^n$. As a first step we will consider only constraints in the form of inequalities

$$\text{Minimize} \quad f(x) \quad \text{subject to} \quad g(x) \leq 0.$$

In this situation we know that optimal solutions must also be solutions of the conditions

$$\nabla f(x) + \mu\,\nabla g(x) = 0, \quad \mu\,g(x) = 0, \quad \mu \geq 0, \quad g(x) \leq 0. \qquad (3\text{--}4)$$

The main result of this section is the following:

Theorem 3.19 *Assume that f and g are convex differentiable functions. If the pair (x, μ) satisfy the KKT conditions above, x is an optimal solution of the problem. If in addition, f is strictly convex, x is the only solution of the problem.*

In fact, this result is almost a consequence of Theorem 3.14, since a solution of the optimality conditions should always be a local minimum. What convexity allows is the passage from a local minimum to a global minimum as stated in that theorem.

A clear, direct proof is almost immediate. Assume that the pair (x, μ) is such that all constraints in (3–4) hold. Notice that the set

$$K = \{x \in \mathbf{R}^n : g(x) \leq 0\}$$

is convex, provided that g is a convex function. This is easy to check. Imagine that y is any other vector in K. We would like to conclude that

$$f(y) - f(x) \geq 0.$$

This is a consequence of the following chain of inequalities, each one of which is explained to the right:

$$
\begin{aligned}
f(y) - f(x) &\geq \nabla f(x)\,(y - x) && \text{(convexity of } f) \\
&= -\mu\,\nabla g(x)(y - x) && \text{(KKT conditions)} \\
&\geq \mu\,(g(x) - g(y)) && \text{(convexity of } g \text{ and } \mu \geq 0) \\
&= -\mu\,g(y) && (\mu\,g(x) = 0) \\
&\geq 0 && (\mu \geq 0, g(y) \leq 0),
\end{aligned}
$$

as desired.

The uniqueness fact is also straightforward if we realize that if $f(x) = f(y)$ for some $x, y \in K$, then all above inequalities are indeed equalities. In particular, we must have

$$f(y) - f(x) = \nabla f(x)\,(y - x)$$

and this implies, by the strict convexity of f, that $x = y$.

This result clearly justifies again the great importance of convexity in minimization problems.

The case in which the NLPP incorporates constraints in the form of equalities places rigid constraints for the sufficiency of optimality conditions. As a matter of fact, since

$$h(x) = 0 \quad \text{is equivalent to} \quad h(x) \leq 0, -h(x) \leq 0,$$

the convexity condition on both h and $-h$ can occur only if h is linear or affine, so that only this type of equality constraints are permitted.

Corollary 3.20   *Assume that f, g are convex differentiable functions and h is affine. Then, optimal solutions for the corresponding NLPP are exactly the solutions of the KKT conditions.*

We will end this section by looking at several examples.

Example 3.21   *We would like to write down optimality conditions for the LPP*

$$\text{Minimize} \quad cx \quad \text{under} \quad Ax = b, x \geq 0.$$

It is immediate to obtain the corresponding KKT conditions

$$c + \lambda A + \mu = 0,$$
$$\mu x = 0,$$
$$Ax = b,$$
$$x \geq 0, \quad \mu \leq 0.$$

By eliminating μ from the first equation we arrive at

$$Ax = b,$$
$$\lambda A + c \geq 0, \quad x \geq 0, \quad (\lambda A + c)x = 0.$$

Since in a LPP all functions involved are linear, they are in particular convex, and therefore any pair (x, λ) verifying the previous restrictions will be an optimal solution of the problem. Notice that the previous system is equivalent to

$$Ax = b, \quad ((\lambda A)_i + c_i)x_i = 0, \quad i = 1, 2, \ldots, n,$$
$$\lambda A + c \geq 0, \quad x \geq 0.$$

Example 3.22   *Find an optimal solution of*

$$\text{Minimize} \quad |x|^2 \quad \text{subject to} \quad ax = c,$$

where a is a given vector, and c is a constant. Since the objective function is strictly convex, and the one in the constraint is linear, we know that the optimal solution (it must be unique if it exists) of this problem corresponds exactly with the unique solution of the KKT conditions. These are

$$2x + \lambda a = 0, \quad ax = c.$$

The unique solution is

$$x = \frac{c}{|a|^2} a, \quad \lambda = \frac{-2c}{|a|^2},$$

so that the minimum value is $c^2 / |a|^2$. Notice that we are calculating the square of the minimum distance to the origin from the hyperplane $ax = c$. Our result coincides with the formula given in elementary analytic geometry, $c/ |a|$.

Example 3.23   *Consider now*

$$\text{Minimize} \quad |x|^2 \quad \text{under} \quad ax \le c, \quad bx \le d,$$

with $a, b, x \in \mathbf{R}^n$, $c, d \in \mathbf{R}$. Once again the objective function is strictly convex, and those involved in the constraints are linear, so that the problem has at most one solution, which, in case it actually exists, must be the only solution of the KKT conditions. These are

$$2x + \mu_1 a + \mu_2 b = 0,$$
$$\mu_1(a\, x - c) = 0, \quad \mu_2(b\, x - d) = 0,$$

together with $\mu_1, \mu_2 \ge 0$, $ax \le c$, $bx \le d$. A full discussion of these equations will lead to four possibilities:

1. $\mu_1 = \mu_2 = 0$, $x = 0$: this will be the optimal solution, provided that $x = 0$ is feasible, i.e., $c \ge 0$ and $d \ge 0$.

2. $\mu_1 = 0$, $\mu_2 = -2d/ |b|^2$, $x = \left(d/ |b|^2 \right) b$: d must be negative, and feasibility of this vector x implies the further restriction

$$dab \le c |b|^2 \,;$$

3. $\mu_2 = 0$, $\mu_1 = -2c/ |a|^2$, $x = \left(c/ |a|^2 \right) a$: c must be negative, and feasibility for x implies

$$cab \le d |a|^2 \,;$$

4. when both multipliers do not vanish, they can be determined as the solution of the linear system

$$|a|^2 \mu_1 + ab\mu_2 = -2c,$$
$$ab\mu_1 + |b|^2 \mu_2 = -2d,$$

whose determinant, $|a|^2 |b|^2 - (ab)^2$, does not vanish unless a and b are collinear. The solution is given by

$$\mu_1 = \frac{2(d\ ab - c\,|b|^2)}{|a|^2\,|b|^2 - (ab)^2}, \quad \mu_2 = \frac{2(c\ ab - d\,|a|^2)}{|a|^2\,|b|^2 - (ab)^2},$$

and the optimal vector is

$$x = -\frac{1}{2}(\mu_1 a + \mu_2 b).$$

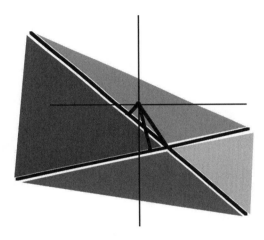

Figure 3.3. The four possibilities in Example 3.23.

To sum up, and depending on the particular data a, b, c, d, we can have the following four situations:

1. $c \geq 0,\ d \geq 0$;
2. $d < 0,\ dab \leq c\,|b|^2$;
3. $c < 0,\ cab \leq d\,|a|^2$;
4. $dab > c\,|b|^2,\ cab > d\,|a|^2$.

It is important to point out that these are not four different solutions but a unique one that depends on the relationship among the different vectors a and b, and scalars c and d. All these possibilities are drawn in Figure 3.3.

Example 3.24   *Consider the problem of finding the minimum of*

$$|x|^4 + |x - a|^2$$

under the constraint

$$|x|^2 \leq 1,$$

where a is a given vector. Due to the strict convexity of the cost function, the optimal solution must correspond to the unique solution of the KKT conditions

$$4|x|^2 x + 2(x - a) + \mu 2x = 0, \quad \mu(|x|^2 - 1) = 0,$$

where we must bear in mind the additional restrictions $\mu \geq 0$, $|x| \leq 1$. The two admissible cases are

$$\mu = 0, \quad x = ta, \quad 2|a|^2 t^3 + t - 1 = 0,$$

and

$$\mu = |a| - 3, \quad x = a/|a|.$$

In the first situation, we must demand $|a| \leq 3$ if $x = ta$ is to be feasible, since

$$\left(2|x|^2 + a\right) x = a.$$

Notice that the cubic polynomial specifying t has a unique real root that lies in the interval $(0, 1/|a|)$ if $|a| \leq 3$, whereas if $|a| \geq 3$, the optimal solution corresponds to the second alternative.

Example 3.25   *A typical truss structure as shown in Figure 3.4 is to be designed according to the criterion of minimum weight subject to a constraint on the maximum deflection permissible at the free node and a lower bound on the crossectional areas of members. The data of the problem are*

$$a_1, a_2, A_0, A_1, A_2, x_0.$$

They should all be positive and depend upon the geometry of the truss, material constants, loads at the indicated points, etc. Specifically, the problem can be stated as

$$\text{Minimize} \quad a_1 x_1 + a_2 x_2$$

subject to

$$\frac{A_1}{x_1} + \frac{A_2}{x_2} \le A_0, \quad x_1, x_2 \ge x_0,$$

where x_1 and x_2 are precisely the crossectional areas to be designed.

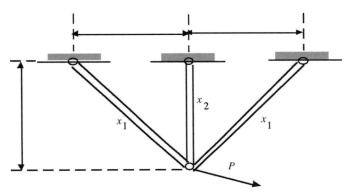

Figure 3.4. A truss structure.

The reader is invited to check that this NLPP is convex, so that the optimal solution can be found by solving the KKT conditions. Namely, if μ_i, $i = 1, 2, 3$, are the multipliers associated with the three restrictions in the form of inequalities, we have

$$a_1 - \frac{\mu_1 A_1}{x_1^2} - \mu_2 = 0,$$

$$a_2 - \frac{\mu_2 A_2}{x_2^2} - \mu_3 = 0,$$

$$\mu_1 \left(\frac{A_1}{x_1} + \frac{A_2}{x_2} - A_0 \right) = 0,$$

$$\mu_2(x_0 - x_1) = 0,$$

$$\mu_3(x_0 - x_3) = 0,$$

$$\frac{A_1}{x_1} + \frac{A_2}{x_2} - A_0 \le 0,$$

$$x_0 - x_1 \le 0, \quad x_0 - x_2 \le 0,$$

$$\mu_1 \ge 0, \quad \mu_2 \ge 0, \quad \mu_3 \ge 0.$$

A full discussion of the solution would require a number of different cases depending on the particular current values of the data set above. For definiteness, we will take

$$a_1 = \frac{1}{5}, \quad a_2 = \frac{1}{6},$$
$$A_0 = 12, \quad A_1 = 25, \quad A_2 = 100,$$
$$x_0 = 10,$$

all given in appropriate units. For this particular data set, the optimal solution turns out to be

$$x_1 = 10, \quad x_2 = \sqrt{120},$$

with multipliers

$$\mu_1 = \mu_3 = 0, \quad \mu_2 = \frac{1}{5}.$$

Details are left to the interested reader.

6. DUALITY AND CONVEXITY

As in LP, we can associate with every NLPP another NLPP, called its dual, such that there is a close relationship between the two. Since we now feel that NLP is much more complicated than its linear counterpart, duality in NLP is also much more delicate. This section intends to be a mere introduction to the subject. From the practical point of view, duality in NLP appears to be a powerful tool in trying to better approximate optimal solutions in NLP. As such, it is closely connected to convexity, as we will see.

Definition 3.26   Given a primal problem

$$\text{Minimize} \quad f(x) \quad \text{under} \quad g(x) \le 0, \quad h(x) = 0,$$

we define its dual as the NLPP

$$\text{Maximize} \quad \theta(\mu, \lambda) \quad \text{under} \quad \mu \ge 0,$$

where the so-called dual function θ is defined on pairs of multipliers (μ, λ) by putting

$$\theta(\mu, \lambda) = \inf_x \left[f(x) + \mu\, g(x) + \lambda\, h(x) \right].$$

Why the dual problem is defined in this way will become clearer as we proceed to better understand the connection between these two NLPP and link them to the KKT optimality conditions. In a sense, the undelying idea is to incorporate necessary conditions of optimality as part of feasibility for a new problem as follows:

$$\text{Minimize}\quad F(x,\mu,\lambda) = f(x)$$

subject to

$$g(x) \le 0,\quad h(x) = 0,$$
$$\nabla f(x) + \mu\,\nabla g(x) + \lambda\,\nabla h(x) = 0,$$
$$\mu \ge 0,\quad \mu g(x) = 0.$$

The function appearing in the definition of the dual function is known as the Lagrangian associated with the problem

$$L(x,\mu,\lambda) = f(x) + \mu\,g(x) + \lambda\,h(x).$$

Lemma 3.27   *Assume that the functions f, g, and h are such that the infimum defining the dual function θ is always attained for all pairs (μ,λ), $\mu \ge 0$. Let $X = X(\mu,\lambda)$ denote one such point where that infimum is taken on, so that*

$$\theta(\mu,\lambda) = f(X) + \mu\,g(X) + \lambda\,h(X).$$

Then if the function $X(\mu,\lambda)$ is differentiable, so is θ, and

$$\nabla_\mu\theta(\mu,\lambda) = g(X),$$
$$\nabla_\lambda\theta(\mu,\lambda) = h(X).$$

Our justification consists of a straightforward computation. If

$$\theta(\mu,\lambda) = f(X) + \mu\,g(X) + \lambda\,h(X),$$

on the one hand, by the chain rule,

$$\nabla_\mu\theta = (\nabla f(X) + \mu\,\nabla g(X) + \lambda\,\nabla h(X))\,\nabla_\mu X + g(X);$$

but on the other, if the Lagrangian attains its minimum at X, its gradient with respect to x must vanish,

$$\nabla f(X) + \mu\,\nabla g(X) + \lambda\,\nabla h(X) = 0,$$

so that

$$\nabla_\mu \theta = g(X),$$

as desired. We have a similar result with the gradient with respect to λ.

As we argued in the LP case, duality is shown in two steps. The next proposition is typically known as weak duality.

Proposition 3.28   Let f, g, and h be differentiable.
1. We always have

$$\max\left\{\theta(\mu, \lambda) : \mu \geq 0\right\} \leq \min\left\{f(x) : g(x) \leq 0, h(x) = 0\right\}.$$

2. If (μ, λ) is feasible for the dual problem $(\mu \geq 0)$, x is feasible for the primal $(g(x) \leq 0, h(x) = 0)$, and

$$\theta(\mu, \lambda) = f(x),$$

then (μ, λ) and x are optimal solutions for the dual and primal, respectively.

The explanation is elementary. Notice that if $\mu \geq 0$, $g(x) \leq 0$, and $h(x) = 0$, then

$$\theta(\mu, \lambda) \leq f(x) + \mu \, g(x) + \lambda \, h(x) \leq f(x).$$

This implies weak duality. The second part of the statement is also straightforward.

The difference

$$\min\left\{f(x) : g(x) \leq 0, h(x) = 0\right\} - \max\left\{\theta(\mu, \lambda) : \mu \geq 0\right\}$$

is called the duality gap. When there is no such gap, both problems are equivalent, and the primal problem can be solved by means of the dual. This is the main idea of all numerical algorithms to compute optimal solutions by looking at the dual. Apart from the interpretation of the dual problem itself, this is the main reason why the dual problem is important in NLP. Convexity is again the main hypothesis under which the duality gap vanishes.

Theorem 3.29   Assume that f and g are convex differentiable functions, h is affine, and the optimization problem defining the dual function is always solvable. Then both problems, the primal and the dual, are solvable simultaneously and there is no duality gap.

For a justification, let us identify by (P) and (D), the primal and dual problems, respectively. Assume first that the primal is solvable, so that there exist a vector x and multipliers (μ, λ) satisfying the KKT conditions, namely,

$$\nabla f(x) + \mu \nabla g(x) + \lambda \nabla h(x) = 0,$$
$$\mu g(x) = 0, \quad \mu \geq 0, \quad g(x) \leq 0, \quad h(x) = 0.$$

All of these conditions imply that x is feasible for (P), and (μ, λ) is feasible for (D); under the convexity of f and g, and the linearity of h, x is a point of attainment of the minimum for the Lagrangian, but since $\mu \ g(x) = h(x) = 0$, we have

$$\theta(\mu, \lambda) = f(x).$$

Proposition 3.28 implies that (μ, λ) is an optimal solution for (D).

Conversely, assume that (μ, λ) is an optimal solution for the dual. If we then apply the KKT conditions to this NLPP, we obtain

$$\nabla_\mu \theta(\mu, \lambda) - y = 0, \quad \nabla_\lambda \theta(\mu, \lambda) = 0,$$
$$\mu \geq 0, \quad y \geq 0, \quad y\mu = 0,$$

where y is the multiplier associated with the constraint $\mu \geq 0$. Keeping in mind Lemma 3.27, and if x is a point where

$$\theta(\mu, \lambda) = f(x) + \mu g(x) + \lambda h(x),$$

so that

$$\nabla f(x) + \mu \nabla g(x) + \lambda \nabla h(x) = 0,$$

we can reinterpret those optimality conditions as

$$g(x) = \nabla_\mu \theta(\mu, \lambda) = y \geq 0,$$
$$h(x) = \nabla_\lambda \theta(\mu, \lambda) = 0,$$
$$\mu g(x) = \mu y = 0.$$

Under convexity assumptions on f, g and linearity on h, satisfying the KKT conditions ensures that x is an optimal solution for (P).

It is relevant to emphasize that this duality fact means that the minimum

$$\min \left\{ \max \left\{ f(x) + \mu \ g(x) + \lambda \ h(x) : \mu \geq 0 \right\} : g(x) \leq 0, h(x) = 0 \right\}$$

and the maximum

$$\max \{\min \{f(x) + \mu \; g(x) + \lambda \; h(x) : g(x) \le 0, h(x) = 0\} : \mu \ge 0\}$$

are equal. Duality is always a question about whether the min–max operation is reversible.

We end this section by computing the dual function in one particular example.

Example 3.30   *Consider the NLPP*

$$\text{Minimize} \quad x_1^2 + x_2^2 + x_3^2$$

subject to

$$x_1^2 + x_2^2 + 3x_3 \le -\frac{5}{2}, \quad x_1 + x_2 + x_3 = -2.$$

Since in this situation all the convexity requirements hold, finding the dual function for this problem $\theta(\mu, \lambda)$ amounts to solving the KKT conditions, for fixed (μ, λ), forgetting about constraints, i.e., solving the system

$$2x_1 + 2x_1\mu + \lambda = 0,$$
$$2x_2 + 2x_2\mu + \lambda = 0,$$
$$2x_3 + 3\mu + \lambda = 0.$$

The unique solution is easily found to be

$$x_1 = x_2 = \frac{-\lambda}{2(1+\mu)}, \quad x_3 = \frac{-1}{2}(\lambda + 3\mu).$$

Taking these values into the corresponding Lagrangian, we have the dual function

$$\theta(\mu, \lambda) = -\frac{\lambda^2}{2(1+\mu)} - \frac{1}{4}(\lambda + 3\mu)^2 + \frac{5}{2}\mu + 2\lambda.$$

It is now a question of some careful computations to check that the optimal solution for the primal and the dual are linked through duality and the KKT conditions. These optimal solutions are

$$x_1 = x_2 = -\frac{1}{2}, x_3 = -1, \quad \mu = \frac{1}{4}, \lambda = \frac{5}{4},$$

and the optimal value for the primal and the dual is the common value 3/2.

7. EXERCISES

1. Determine the critical points of the function

$$f(x_1, x_2) = (2 - x_1 - x_2)^2 + (1 + x_1 + x_2 - x_1 x_2)^2.$$

Try to decide their nature.
2. Find the minimum of the function

$$\sigma^2 = \sum_{i=1}^{n} T_i^2 x_i^2$$

with respect to the variables x_i subject to

$$c = \sum_{i=1}^{n} x_i.$$

Both c and T_i are fixed constants.
3. Given the objective function

$$P = (x_1 - 1)^2 + x_n^2 + \sum_{i=1}^{n-1} (x_{k+1} - x_k)^2,$$

find the critical points of P
    1. without any constraint;
    2. subject to

$$c = \sum_{i=1}^{n} a^i x_i,$$

where c and a are constants.
4. Check that the function

$$f(x_1, x_2, x_3) = x_1^2 + x_2^2 + x_3^2 - x_1 - x_2 - x_3$$

is convex. Find the extreme values of f under the conditions

$$x_1^2 + x_2^2 = 4, \quad -1 \le x_3 \le 1.$$

5. Describe the region of the plane determined by the two inequalities

$$x_1^2 - x_2^2 \le 1, \quad x_1^2 + x_2^2 \le 4.$$

Find the extreme values of

$$f(x_1, x_2) = x_1^2 + 2x_2^2 + x_1 x_2$$

over that region.

6. Define the functions

$$F(a) = \min \{ f(x_1, x_2, x_3) : g(x_1, x_2, x_3) = a \},$$
$$G(a) = \min \{ f(x_1, x_2, x_3) : g(x_1, x_2, x_3) \le a \},$$

for $a \in \mathbf{R}$ and

$$f(x_1, x_2, x_3) = x_1^4 + x_1^2(1 - 2x_2^2) + x_2^2 + x_3^2 - 2x_1 + 1,$$
$$g(x_1, x_2, x_3) = x_1^4 + x_2^4 + x_3^4.$$

Find explicit expressions for $F(a)$ and $G(a)$. Study the two problems

$$\min F(a), \quad \min G(a)$$

and their relationship. What can you conclude about the minimum of the function f over all of $\mathbf{R}^3$? Do the same for simpler choices of the function g.

7. Find the maximum and minimum values of

$$f(x_1, x_2) = \int_{x_1}^{x_2} \frac{1}{1 + t^4} \, dt$$

over the region determined by $x_1^2 x_2^2 = 1$.

8. Find the closest point of the surface $xy + xz + yz = 1$ to the origin. Do the same for the surface of equation $x^2 + y^2 - z^2 = 1$.

9. Solve the problem of the Cobb–Douglas utility function of Chapter 1.

10. Solve the problem of the location of several service points where clients are known (Chapter 1) for the following data set: The locations of customers are

$$(1,0), (2,1), (-1,2), (3,-1), (-1,-2), (3,-2),$$

and three service points are to be built.

11. Solve the exercise of the ladder in Chapter 1.

12. Solve the problem of the scaffolding system of Chapter 1.

13. A certain set of experimental data relating two variables x and y is at our disposal,

$$(x_i, y_i), \quad i = 1, 2, \ldots, n.$$

A linear relationship between x and y is desired, but this is typically not possible in an exact fashion. Determine the best coefficients a, b such that the quadratic error of the data set with respect to the desired linear model

$$y = ax + b$$

is minimized.

14. When a certain linear system $Ax = b$ is not solvable, we might yet be interested in the vector x "closest" to being a solution by minimizing the quadratic (or other type of) error

$$\text{Minimize} \quad \frac{1}{2} |Ax - b|^2$$

over all possible x. Solve this problem in general and apply your solution to the particular case

$$\begin{pmatrix} 1 & -1 \\ 1 & 1 \\ 2 & -1 \end{pmatrix} \begin{pmatrix} x_1 \\ x_2 \end{pmatrix} = \begin{pmatrix} 4 \\ 0 \\ 0 \end{pmatrix}.$$

15. A truncated-conic bar, clamped at its upper end and hanging vertically (in the spirit of Exercise 10 of Chapter 1) is to be designed under the following restrictions (Figure 3.5): total length L, total volume at our disposal V, density of material to be used ρ, Young modulus of material to be used E. The criterion is to minimize total elongation under the action of a given weight W at its lower tip and its own weight. We assume that Hooke's law is valid: If the cross section at distance x from the upper end moves to $y(x)$

under the action of W, then the strain at such section $y'(x)$ is proportional (with constant $1/E$) to the stress in such a section, and this stress in turn is the quotient of the total load acting on it and the corresponding area of the cross section. The radii R and r of the upper and lower sections are to be determined.

If the cross sections are squares, so that the bar is a truncated pyramid, is the same optimal solution expected?

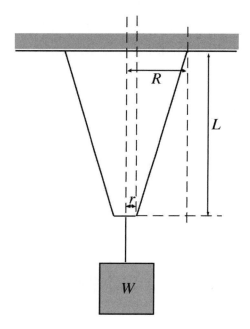

Figure 3.5. A conic bar.

Chapter 4

Approximation Techniques

1. INTRODUCTION

It is likely that our readers may have already realized that solving optimization problems explicitly is not an easy task. As a matter of fact, it is typically an impossible job. Not only for those problems with a high number of variables involved is it virtually hopeless to compute by hand the optimal solutions, but even for many modest-sized problems it is almost impossible to solve and manipulate so many equations. It is therefore of primary importance to show how solutions for optimization problems can be efficiently approximated. This need is even more unavoidable from the engineering and practical point of view, since explicit, accurate approximation is as important as the understanding of the underlying problem. As usual, our aim in this chapter is to cover the basic algorithms that researchers have developed over the years to approximate solutions for NLPP without trying to exhaust all possibilities, describe the most recent trends, or even show where the algorithms come from and why they have their particular structure. We will try to motivate, however, the most popular ones so that the reader may have a feeling of their nature without entering into technical details. It is also true that this is a highly technical subject evolving very rapidly, so that the methods that seem best now will probably be abandoned in a few years and replaced either by old ideas in a new framework or by entirely novel techniques. See, for instance, [17] for a very nice survey on all this and the importance of interior point methods nowadays.

There is a further, important, reason why we will not describe a full list of algorithms or go into too many details, and that is that there already exist very powerful software packages devoted to approximate optimal solutions in a variety of situations. These tools free the user from the need to be too concerned about technicalities related to practical algorithms and focus on the modeling issues of problems and the interpretation and assessment of their solutions. Some of these software packages are AIMMS, AMPL, AMSL, GAMS, Optimization Toolbox of MatLab, and SNOPT. Much information on all this is scattered throughout the Internet. We especially recommend the site [29], which is like a master site for optimization. In particular, when one is looking for information on what software is best for one's needs, this site is a must to visit.

Those readers interested in deepening their understanding of approximation techniques and willing to pursue this direction should definitely resort to some of the references we have selected at the end of the text. In particular, algorithms for large optimization problems require considerable expertise, patience, and study. Approximation is much more than the content of this chapter. For more comprehensive sources, see [1], [13], [16], [22], [30], [32].

Many numerical approximation methods for NLPP problems have an iterative nature. This means that the approximation scheme proceeds in successive, better approximations to the solutions sought. In this way, any such algorithm must specify a mechanism to build a new iteration from one (or several) we already have. We thus construct a sequence $\{x_k\}$ of successive approximations to the real optimal solution x, trusting that x_{k+1} will be closer to x than it was in its preceding approximation x_k. Indeed, an exponent $\alpha > 0$ characterizes efficient numerical algorithms when

$$|x_{k+1} - x| \le C \, |x_k - x|^\alpha, \quad C > 0.$$

If we interpret the difference

$$|x_j - x|$$

as a measure of the error we make when taking x_j for x, the above inequality says that in each new iteration the error is reduced by a power with exponent α. The bigger α is, the better the method, since the error is decreased more rapidly, but the more expensive the computation of the new iteration may be. There should always be a balance between the efficiency of the algorithm and the computational cost associated with its implementation.

Each specific numerical algorithm should precisely determine the passage from one iteration x_k to the next x_{k+1}. Since these are vectors, we can think of this process as a two-step strategy:

1. **Search direction**: Decide on the vector d_k pointing toward x_{k+1} from x_k, such that d_k is parallel to $x_{k+1} - x_k$; this vector d_k is called a search direction.
2. **Step-size parameter**: Once d_k has been decided upon, try to determine a parameter t_k such that

$$x_{k+1} = x_k + t_k d_k;$$

t_k is called the step-size parameter.

These two elements, d_k and t_k, suffice to determine a new iteration from a given one. Each complete optimization algorithm must address and incorporate these two ingredients: search direction and step-size parameter.

We will restrict attention first to NLPP without constraints, and treat in this context some of the algorithms for deciding on the step-size parameter and on the search direction. In the first case, we will briefly describe the fixed variable step-size, interpolation, and golden rule algorithms, and we will focus on the steepest-descent, the conjugate gradient, and Newton-like methods for the second. Later, we will be concerned with the main ideas for dealing with NLPP with constraints including penalizations and barriers, the dual method, and finally, the augmented-Lagrangian method. We will try to remain as specific and to the point in each situation as possible, with the idea of not confusing our readers with too many statements.

2. LINE SEARCH METHODS

We will start by treating the numerical approximation of the problem

$$\text{Minimize} \quad g(x), \qquad x \in \mathbf{R}^n.$$

As pointed out in the previous section, the choice of the step size must be made after a search direction has been decided upon. Therefore, we will assume that a search direction determined by d_k emanating from the approximation x_k has been chosen, and we would like to decide on the parameter t_k such that

$$x_{k+1} = x_k + t_k d_k$$

will be our next iterate to the optimal solution we seek. If we are interested in finding the global minimum of $g(x)$, where we assume g to be defined in all of $\mathbf{R}^n$, and the corresponding minimum problem does have a solution, the best move we can do toward that goal is to choose t_k as a solution of the problem

$$\text{Minimize} \quad g(x_k + td_k), \qquad t \in \mathbf{R}.$$

If we put

$$h(t) = g(x_k + td_k), \quad t \in \mathbf{R},$$

we are led to consider the one-dimensional minimization problem on the function h. Since h is a one-dimensional section of g, this step is typically referred to as a line search scheme. Hence, we concentrate on finding an approximation for

$$\text{Minimize} \quad h(t), \qquad t \in \mathbf{R}.$$

We proceed to describe briefly some of the simpler methods that can be and are used for the one-dimensional situation. We could certainly try to determine exactly the minimum point for the function h, but again this strategy is not appealing from the practical point of view, since even if could exactly find that minimum point, the fact that we will have to solve such a situation many, many times in a systematic fashion for higherdimensional problems points toward finding an efficient way of approximating the value of the minimum point for one-dimensional problems.

Fixed or variable step size. *This is the most elementary of all methods. It consists in selecting a fixed step size t and setting successive approximations by putting*

$$t_{i+1} - t_i = t.$$

We proceed in this fashion until

$$h(t_{i+1}) \geq h(t_i);$$

at this point we reduce the size of t by a significant factor, and we change its sign. We now proceed by taking

$$t_{i+1} - t_i = t$$

with this new value of t until again we get

$$h(t_{i+1}) \geq h(t_i).$$

We keep repeating this process iteratively until a preassigned precision is reached.

Interpolation. The interpolation method consists in interpolating the values of h at three points t_1, t_2, t_3, by a quadratic polynomial whose minimum point is easily located. It is convenient for the relative location of the values of h at these three points to be

$$h(t_2) < h(t_1), \quad h(t_2) < h(t_3).$$

Such a point of minimum, t^*, for the interpolating polynomial is considered a good approximation of the real point of minimum for h. If more precision is desired, t^* replaces one of the t_i's so that the relative position of the values of h at the three chosen points is as above, and computations are redone until a prescribed threshold precision is hit. For smooth functions the method turns out to be quite efficient, specially because fully explicit formulas for t^* in terms of t_i and $h(t_i)$ can be written down. Indeed,

$$t^* = \frac{1}{2} \frac{t_1^2(h(t_2) - h(t_3)) + t_2^2(h(t_3) - h(t_1)) + t_3^2(h(t_1) - h(t_2))}{t_1(h(t_2) - h(t_3)) + t_2(h(t_3) - h(t_1)) + t_3(h(t_1) - h(t_2))}.$$

Notice that the method is exact with a single iteration when h is quadratic.

Golden section method. The golden section algorithm tries to reduce the size of the interval where the minimum point is located by a factor $k = 0.618034$. This factor is the golden section, which is the positive solution of the equation $1 + k = 1/k$. Let us suppose that the minimum is attained at a point in the interval (t_1, t_2). We consider two points

$$t_3 = kt_1 + (1 - k)t_2, \quad t_4 = kt_2 + (1 - k)t_1.$$

Comparing the values of $h(t_3)$ and $h(t_4)$ we proceed as follows:
1. *if $h(t_3) > h(t_4)$, the minimum is located in the interval (t_3, t_2), so that t_3 replaces t_1;*
2. *if $h(t_3) < h(t_4)$, then the minimum belongs to the inteval (t_1, t_4), and t_4 takes the place of t_2.*

We apply this procedure iteratively until the interval has a length less than a preassigned threshold value.

Fibonacci's method. *This is a variant of the previous one in which the ratio k varies depending on the iteration we are computing. The values of k change according to the Fibonacci sequence, which is defined recursively by*

$$a_1 = a_2 = 1, \quad a_{j+2} = a_{j+1} + a_j, \quad j \geq 1.$$

In each iteration, we use the parameter

$$k_j = \frac{a_j}{a_{j+1}}$$

instead of k. Notice that k_j tends to k, the golden section, as $j \to \infty$.

It is not fair to say that a particular method is always better than any other one. Depending on the problem, one method may be preferable over other possibilities. Each user may find his/her own preferences by experimentation. In general, we can say that the interpolation method works quite well when the objective function is smooth. Otherwise, the golden section scheme may be used. The fixed step-size algorithm can be used, however, as a generalpurpose algorithm to be utilized in any situation. There exist other more sophisticated schemes. See the references cited in the introduction to this chapter.

3. GRADIENT METHODS

Once we have treated the issue of the step-size choice when a search direction has been determined, we study the task of deciding on this direction. This a more complex and fundamental question. The success of a particular method is greatly influenced by a good algorithm for choosing these search directions. There are essentially two groups of algorithms to decide on the search direction:

those not requiring any information on the objective function, and those based on the information provided by the gradient of the objective function. Since it is reasonable to believe that we can use the information coming from the gradient to our advantage, we will focus on the gradient methods, since these are extensively used. We will describe three types of methods that are among the most popular choices: the steepest descent, conjugate gradient, and quasi-Newton.

The basic idea of all these algorithms relies on the notion of a descent direction for a given smooth function.

Definition 4.1 *(Descent direction) A vector $d \in \mathbf{R}^n$ is a descent direction for a smooth function $g : \mathbf{R}^n \to \mathbf{R}$ at a point $x \in \mathbf{R}^n$ if*

$$\nabla f(x)d < 0.$$

The reason for this definition is quite simple. Just notice that if we define

$$h(t) = g(x + td),$$

then by the chain rule,

$$h'(0) = \nabla f(x)d.$$

Therefore, if d is a descent direction, this derivative is negative, and hence the values of g decrease as we move along d from x, at least locally.

Any gradient method is associated with different ways of choosing the descent direction. As announced, conjugate gradient and quasi-Newton methods are among the most popular and efficient candidates. We will briefly describe the steepest descent and quasi-Newton methods, but dwell a bit more on conjugate gradient methods.

Steepest descent. *It is well known that the gradient of a function at a point, $\nabla g(x)$, provides the direction along which the function increases most rapidly. Consequently, $-\nabla g(x)$ furnishes the direction of "steepest descent" from the point x. Thus at each iteration x_k, the gradient of g is evaluated at x_k, and we take as search direction*

$$d_k = -\nabla g(x_k).$$

Notice that d_k is always a descent direction, indeed the steepest descent direction, since

$$\nabla g(x_k)d_k = -\|\nabla g(x_k)\|^2 \leq 0.$$

If the gradient vanishes, then x_k is the minimum point sought. If it does not then d_k is a descent direction. In summary, the steepest descent algorithm can be described as follows:

1. **Initialization.** Choose x_0, initial approximation.
2. **Search direction.** Given approximation x_k, set $d_k = -\nabla g(x_k)$.
3. **Stopping criterion.** For a preassigned threshold value $\epsilon > 0$, if

$$\|d_k\| \leq \epsilon,$$

stop: The current approximation x_k is sufficiently good, i.e., sufficiently close to a true point of minimum. Else, continue. Notice that at such a point of minimum the gradient actually vanishes.

4. **Line search.** Find an approximation to the line minimization problem (see last section)

$$\text{Minimize} \quad g(x_k + td_k), \qquad t \in \mathbf{R}.$$

Let t_k be an approximation to such a minimum point.

5. **New approximation.** Put $x_{k+1} = x_k + t_k d_k$. Go back to step 2.

Quasi-Newton methods   The Newton method takes as search direction

$$d = -[\nabla^2 g(x)]^{-1} \nabla g(x),$$

where

$$\nabla^2 g(x)$$

is the (symmetric) Hessian matrix of g at x.

Proposition 4.2   If $\nabla^2 g(x)$ is positive definite, then d as above is a descent direction for g at x. Indeed, if A is any positive definite matrix, the direction

$$-A\nabla g(x)$$

is a descent direction of g at x.

The proof is elementary and left to the reader. More important than this fact is to understand where the descent direction for Newton's method comes from. From Taylor's expansion we have

$$g(x) \approx g(x_0) + \nabla g(x_0) \ (x - x_0) + \frac{1}{2}(x - x_0)^T \nabla^2 g(x_0)(x - x_0),$$

so that by differentiation,

$$\nabla g(x) \approx \nabla g(x_0) + \nabla^2 g(x_0) \ (x - x_0).$$

If x is a point of minimum, we must enforce $\nabla g(x) = 0$, and this leads to

$$x = x_0 - [\nabla^2 g(x_0)]^{-1} \nabla g(x_0).$$

This is the explanation of the form of the search direction for Newton's method.

The computation of the inverse of the Hessian matrix is, however, not appealing from a practical point of view. Trying to overcome this difficulty leads one to consider quasi-Newton algorithms where the inverse of the Hessian matrix is approximated succesively by using only the gradient of g (first derivatives). We simply give in the sequel the two most important quasi-Newton algorithms without further justification.

1. **Initialization.** x_1, *initial approximation;* $H_1 = \mathbf{1}$, $p_1 = -H_1 \nabla g(x_1)$. *Here $\mathbf{1}$ is the identity matrix.*
2. **Line search.** *Find an approximation to the line minimization problem*

$$Minimize \quad g(x_k + tp_k), \qquad t \in \mathbf{R}.$$

Let t_k be an approximation to such a minimum point.
3. **New approximation.** *Put $x_{k+1} = x_k + t_k d_k$.*
   3. **Stopping criterion.** *For a preassigned threshold value $\epsilon > 0$, if*

$$\|\nabla g(x_{k+1})\| \leq \epsilon,$$

   stop: The current approximation x_{k+1} is sufficiently good, i.e., sufficiently close to a true point of minimum. Else, continue.
4. **New search direction.** *Let*

$$q_k = \nabla g(x_{k+1}) - \nabla g(x_k), \quad p_{k+1} = -H_{k+1} \nabla g(x_{k+1}),$$

and go to step 2.

The form of the matrix H_{k+1} distinguishes different algorithms.
1. **Davidon–Fletcher–Powell algorithm.** *Take*

$$H_{k+1} = H_k + t_k \frac{p_k p_k^T}{p_k q_k} - \frac{H_k q_k q_k^T H_k}{q_k^T H_k q_k}.$$

2. **Broyden–Goldfarb–Shanno algorithm.** *Take*

$$H_{k+1} = t_k \frac{p_k p_k^T}{q_k p_k} + \left(1 - \frac{p_k q_k^T}{q_k p_k}\right) H_k \left(1 - \frac{q_k p_k^T}{q_k p_k}\right).$$

4. CONJUGATE GRADIENT METHODS

The conjugate gradient algorithm is a more elaborate procedure to decide on the search direction as compared to the steepest descent method. Its interest can be motivated as a way to solve the problem that the direction provided by the gradient itself is often not the best choice, and therefore the steepest descent algorithm may be extremely slow in getting close to the minimum point. The main concept associated with the conjugate gradient algorithm is that of conjugate directions for a quadratic function. In all that follows, we will restrict attention, to motivate the genesis of the algorithm itself, to a quadratic function

$$g(x) = \frac{1}{2} x^T A x - b x + c,$$

where A is a positive definite $n \times n$ matrix (so that the corresponding minimum problem has a unique solution), b is a vector, and c is a constant.

Definition 4.3   *A set of vectors $\{p_i\}$, $i = 1, 2, \ldots, n$, is a set of conjugate directions for g if*

$$p_i^T A p_j = 0, \quad i \neq j, \qquad p_i^T A p_i = \gamma_i, \quad i = 1, 2, \ldots, n.$$

In particular, this set of vectors is linearly independent, and they make up a basis for $\mathbf{R}^n$. We specify the interest of conjugate directions with respect to minimum problems in two results. The first one establishes the relevance of conjugate directions with respect to minimum problems. The second one indicates one of various possibilities for determining sets of conjugate directions for a given quadratic function. We finally explicitly write down the conjugate gradient algorithm for a general objective function g, not necessarily quadratic.

Lemma 4.4   *Let $\{p_j\}$ be a set of conjugate directions with respect to the quadratic function g as above, and put:*

1. x_1 arbitrary, $g_1 = -\nabla g(x_1)$;
2. for $k \geq 1$,

$$t_k = -\frac{g_k p_k}{p_k^T A p_k}, \quad x_{k+1} = x_k + t_k p_k, \quad g_{k+1} = \nabla g(x_{k+1}) = g_k + t_k A p_k.$$

For every j we have

$$g_{j+1} p_k = 0, \quad k = 1, 2, \ldots, j.$$

In particular,

$$g_{n+1} p_k = 0, \quad k = 1, 2, \ldots, n,$$

implies $g_{n+1} = 0$, and consequently, g attains its minimum at x_{n+1}.

This result says that by using a set of conjugate directions as search directions for a quadratic functional, we can find the minimum in exactly n steps, where n is the dimension of the problem. It is interesting to point out where the choice of t_k comes from. If we consider the function

$$h(t) = g(x_k + t p_k),$$

it turns out that the value of t at which the minimum of h is taken on is precisely the value we have chosen for t_k. This is easily derived by keeping in mind that g is quadratic and appropriately using the chain rule.

For the proof of Lemma 4.4, notice that the minimum we are looking for is the solution of the linear system

$$Ax - b = 0,$$

since the gradient of g is

$$\nabla g(x) = Ax - b.$$

Therefore, we pretend to solve the previous linear system in n steps n being the dimension of the matrix. We argue by induction on the index j, so that we assume

$$g_{j+1} p_k = 0, \quad k = 1, 2, \ldots, j,$$

and want to conclude that

$$g_{j+2} p_k = 0, \quad k = 1, 2, \ldots, j + 1.$$

On the one hand,

$$g_{j+2}p_k = \left(g_{j+1} + t_{j+1}^T A p_{j+1}\right) p_k = 0$$

if $k = 1, 2, \ldots, j$, by the induction hypothesis and the fact that we are working with conjugate directions. On the other hand, the identity

$$g_{j+2}p_{j+1} = \left(g_{j+1} + t_{j+1}^T A p_{j+1}\right) \cdot p_{j+1} = 0$$

follows from the choice of t_{j+1}. Notice that the choice of t_k is dictated so as to determine the minimum of the quadratic function of t, $g(x_k + t p_k)$, as remarked before.

Once we have seen the interest of the sets of conjugate directions, we provide and justify one of various important ways of recursively constructing such sets of directions and the succesive approximations to the point of minimum, simultaneously.

Lemma 4.5   (Fletcher–Reeves) Set:
1. $p_1 = -g_1 = -\nabla g(x_1)$;
2. for $k \geq 1$,

$$p_{k+1} = -g_{k+1} + \beta_k p_k, \qquad \beta_k = \frac{|g_{k+1}|^2}{|g_k|^2},$$

where g_j is the gradient of g at x_j. The set $\{p_j\}$ is a set of conjugate directions for g.

To prove this result, we again argue by induction. Suppose we have chosen k conjugate directions p_j, $j = 1, 2, \ldots, k$, so that, according to Lemma 4.4,

$$g_{k+1}p_j = 0, \quad t_j^T A p_j = g_{j+1} - g_j, \quad j = 1, 2, \ldots, k, \tag{4-1}$$

and

$$t_j = -\frac{g_j p_j}{p_j^T A p_j} = -\frac{g_j \left(-g_j + \beta_{j-1} p_{j-1}\right)}{p_j^T A p_j} = \frac{|g_j|^2}{p_j^T A p_j},$$

since $g_j p_{j-1} = 0$. If we take the new direction p_{k+1} as in the statement of the lemma, we would like to conclude that it is conjugate with respect to the previous ones. Bearing in mind (4–1), we have

$$
\begin{aligned}
p_{k+1}^T A p_j &= (-g_{k+1} + \beta_k p_k)^T A p_j \\
&= -g_{k+1} \frac{1}{t_j}(g_{j+1} - g_j) + \beta_k p_k^T A p_j \\
&= -\frac{g_{k+1}}{t_j}(-p_{j+1} + \beta_j p_j + p_j - \beta_{j-1} p_{j-1}) + \beta_k p_k^T A p_j.
\end{aligned}
$$

For $j < k$, and assuming that t_j does not vanish (otherwise, $g_j = 0$, which implies that x_j is the point of minimum, and there is no need for more conjugate directions), we observe that this expression is zero by the induction hypothesis and the fact that the previously chosen directions are conjugate among themselves. For $j = k$, some terms vanish, and others do not. Specifically, having in mind the formula for β_k, we obtain

$$
\begin{aligned}
p_{k+1}^T A p_k &= \frac{g_{k+1} p_{k+1}}{t_k} + \beta_k p_k^T A p_k \\
&= \frac{p_k^T A p_k}{|g_k|^2}\left(-|g_{k+1}|^2 + \beta_k g_{k+1} p_k\right) + \beta_k p_k^T A p_k \\
&= \frac{p_k^T A p_k}{|g_k|^2}\left(-|g_{k+1}|^2 + \frac{|g_{k+1}|^2}{|g_k|^2} g_{k+1} p_k + |g_{k+1}|^2\right).
\end{aligned}
$$

This last expression vanishes because $g_{k+1} p_k = 0$.

For a general objective function $g(x)$ the conjugate gradient method proceeds by successive approximations, and a new search direction is chosen at every iteration. The form of these search directions is justified by the previous discussion. In algorithmic fashion, we can sum up the conjugate gradient method as follows:

1. **Initialization.** Choose x_0, initial approximation.
2. **Stopping criterion.** For a preassigned threshold value $\epsilon > 0$, if

$$
\|\nabla g(x_k)\| \le \epsilon,
$$

   stop: The current approximation x_k is sufficiently good, i.e., sufficiently close to a true point of minimum. Else, continue.
3. **Search direction.** Given approximation x_k, set $p_k = -\nabla g(x_k) + \beta_k p_{k-1}$, taking p_{-1} arbitrary.

4. **Line search**. Find an approximation to the line minimization problem (see Section 4.2)

$$\text{Minimize} \quad g(x_k + tp_k), \qquad t \in \mathbf{R}.$$

Let t_k be an approximation to such minimum point.

5. **New approximation**. Put $x_{k+1} = x_k + t_k d_k$. Go back to step 2.

The different variants of the conjugate gradient algorithm correspond to different ways of taking the parameter β_k. The most popular choices are the following:

1. Fletcher–Reeves algorithm:

$$\beta_k = \begin{cases} 0, & k = jn, \; j = 0, 1, \ldots, \\ \dfrac{\|\nabla g(x_k)\|^2}{\|\nabla g(x_{k-1})\|^2}, & \text{otherwise.} \end{cases}$$

2. Polak–Ribière algorithm:

$$\beta_k = \begin{cases} 0, & k = jn, \; j = 0, 1, \ldots, \\ \dfrac{\nabla g(x_k) \, (\nabla g(x_k) - \nabla g(x_{k-1}))}{\|\nabla g(x_{k-1})\|^2}, & \text{otherwise.} \end{cases}$$

The reason for taking $\beta_k = 0$ every n iterations is to avoid the effect of accumulation of numerical errors. Empirically, the Polak–Ribière algorithm seems more robust.

5. APPROXIMATION UNDER CONSTRAINTS

Since constraints are frequently an essential part of optimization problems, numerical algorithms to solve or approximate solutions must be such that they respect the appropriate restrictions and indeed lead to the solution of the optimization problem subjected to those constraints. We are therefore interested in describing algorithms to efficiently approximate the optimal solutions of a typical NLPP,

$$\text{Minimize} \quad f(x)$$

subject to

$$g(x) \leq 0, \quad h(x) = 0.$$

There are essentially two main strategies for treating this problem numerically: Either we decide not to involve multipliers, and therefore not to use the information coming from optimality conditions; or else, we try to utilize this information in some way. The first class includes the techniques of penalization and barriers, and in the second category we will explain the standard dual method and the augmented Lagrangian algorithm. In any case, algorithms are built in such a way that they rely in one way or another on the nonconstraint case, so that the underlying idea is to construct a closely related, unconstrained optimization problem and apply to it some of the algorithms we already have for nonconstrained problems.

Penalization and barriers. *The idea behind the penalization and barrier methods consists in transforming the optimization problem with constraints in such a way that infeasible vectors are prohibited or at least penalized. Ideally, this can be accomplished by considering the following optimization problem:*

$$\text{Minimize} \quad f(x) + \tilde{f}(x),$$

where

$$\tilde{f}(x) = \begin{cases} 0, & g(x) \le 0, h(x) = 0, \\ +\infty, & otherwise. \end{cases}$$

The effect of adding $\tilde{f}$ to f, as one can see in the previous definition, is null if a vector is feasible, so that its cost is $f(x)$, as it should be. But if x is not feasible, its cost is infinite, so that it is eliminated from the minimization process. This is exactly what the constraints mean. The new problem is a nonconstrained one. However, the fact that the cost $f + \tilde{f}$ is not continuous, since it can take on the value $+\infty$ abruptly, makes this new problem unsuitable for the algorithms we explained in the last section. Before applying such algorithms to the problem, we must do something about this special cost $f + \tilde{f}$.

One possibility is not to assign an infinite cost to an infeasible vector, but simply to penalize, in one way or another, those vectors. This is the idea of the penalization method that is used in many other areas of mathematics: If we ought to satisfy some restrictions that are hard to handle, we can ignore them but add a penalization term to the cost functional when they are not met so as to discourage them. One of the most popular families of penalizations is

$$\tilde{f}_r(x) = r\left(\sum_i \max\left\{0, g_i(x)\right\}^p + \sum_j |h_j(x)|^q\right),$$

where $p, q > 1$ are exponents and r is a penalization parameter. Notice that if $p > 1$, the function $\max\{0, g_i(x)\}^p$ is differentiable if g_i is. The greater r is, the more effective the penalization. One typical case that is used often is the quadratic penalization corresponding to $p = q = 2$. It is true that since a penalization does not prohibit infeasible points, the approximation one can obtain by applying a nonconstrained optimization algorithm to $f + \tilde{f}_r$ may not give good results. Only when we push the parameter r to become bigger and bigger, the corresponding approximations to the nonconstrained problems are closer and closer to one solution of the constrained problem.

Another idea that can be used in dealing with constrained problems is to mimic the infinite barrier that $\tilde{f}$ above sets between feasible and infeasible points, but in such a way that it is not instantaneous. For example, if we take

$$\tilde{f}_r(x) = -\frac{1}{r} \sum_i \frac{1}{g_i(x)},$$

when the constraints of the problem come only in the form of inequalities $g(x) \leq 0$, we notice that as we move toward the boundary of the feasible set so that $g_i(x)$ becomes closer to zero from the negative part, the function $\tilde{f}_r(x)$ becomes larger and larger and eventually takes on an infinite value, placing a barrier at the boundary of the set of feasible points. The effect of the parameter r when it becomes big is to interfere as little as possible with the value of the objective function f on the set of feasible points. When $g_i(x) < 0$ and r is large, so that $1/r$ is small, we have that $\tilde{f}_r(x)$ is also very small. Again, only when this parameter r is large can we obtain good approximations by applying a nonconstrained algorithm to the cost $f + \tilde{f}_r$. As $r \to +\infty$ those approximations will tend to a true solution of the original constrained problem. Another interesting posibility is to take a logarithmic barrier of the type

$$\tilde{f}_r(x) = -\frac{1}{r} \sum_i \log(-g_i(x)).$$

One further good choice can be

$$\tilde{f}_r(x) = \frac{1}{r} \sum_i \frac{1}{1 - rg_i(x)}.$$

In the case in which some constraints come in the form of equalities $h(x) = 0$, one possibility consists in adding the term

$$r^3 \sum_j \frac{h_j(x)^2}{1 - r^2 h_j(x)^2}$$

to $\tilde{f}_r$.

Although the preceding ideas look appealing for their simplicity, in practice, due to the fact that accuracy is linked to high values of the parameter r, numerical errors appear because we have to deal with very large numbers, and only obtains one a modest efficiency. Indeed, optimal results in using penalization or barriers are found in an intermediate range for the parameter r, and this range is highly dependent on the particular problem at hand, so that a good deal of experimentation is required to assess optimal results. For this reason, algorithms taking into account multipliers and optimality and/or duality conditions have been developed. We will restrict attention, as pointed out before, to a standard dual method and will finish with a short discussion of the augmented Lagrangian algorithm.

Dual method.   *We learned in Chapter 3 the relationship between an NLPP and its dual. In fact, if we have to solve the primal problem*

$$\text{Minimize} \quad f(x)$$

subject to

$$g(x) \leq 0, \quad h(x) = 0,$$

we know that under appropriate hypotheses, we can equivalently treat its dual

$$\text{Maximize} \quad \theta(\mu, \lambda)$$

under $\mu \geq 0$, where the dual function is defined by

$$\theta(\mu, \lambda) = \min_x \left\{ f(x) + \mu\, g(x) + \lambda\, h(x) \right\}.$$

The advantage of the dual is that the definition of the dual function is a nonconstrained problem, and, at the same time, the constraint itself for the dual problem is much simpler (in particular, linear) than in the primal problem. The idea of the dual method consists in approximating the optimal solution of the dual, and then computing an approximation of the optimal solution of the primal. For notational convenience, let us put

$$L(x, \mu, \lambda) = f(x) + \mu\, g(x) + \lambda\, h(x)$$

for the Lagrangian of the problem. In the following algorithm we have used a
steepest descent method for the solution of the dual. Remember (Chapter 3)
that
$$\nabla_\mu \theta(\mu, \lambda) = g(x), \quad \nabla_\lambda \theta(\mu, \lambda) = h(x),$$
if x is such that $\theta(\mu, \lambda) = L(x, \mu, \lambda)$.

1. **Initialization.** (μ_1, λ_1), initial approximation with $\mu_1 > 0$.
2. **Approximated solution for the primal.** For (μ_j, λ_j), find (approximate) an
   optimal x_j for the dual function. Use a nonconstrained algorithm for this.
3. **Stopping criterion.** For a preassigned threshold value $\epsilon > 0$, if
$$|h(x_j)| < \epsilon, \quad |\mu_j \, g(x_j)| < \epsilon,$$
   stop: The current approximation x_j is sufficiently good. Else continue.
4. **Search direction.** Take
$$d_k = \begin{cases} g_k(x_j), & \text{if } \mu_j^{(k)} > 0, \\ \max\{0, g_k(x_j)\}, & \text{if } \mu_j^{(k)} = 0, \end{cases}$$
   where $\mu^{(k)}$ represents the k-component of μ, and
$$e_k = h_k(x_j).$$
5. **New approximation.** Put
$$\mu_{j+1} = \mu_j + s_j d, \quad \lambda_{j+1} = \lambda_j + s_j e,$$
   where s_j is chosen to maximize the function
$$\varphi(s) = \theta(\mu_j + sd, \lambda_j + se),$$
   keeping in mind that s is restricted because $\mu_j + sd \geq 0$. Go to step 2.

Augmented Lagrangian method. Another fruitful idea consists in using
the information coming from optimality conditions. In order to motivate the
form of the final algorithm we will present, let us focus first on an NLPP with
equality constraints of the type

$$\text{Minimize} \quad f(x) \quad \text{under} \quad h(x) = 0.$$

We know that optimal solutions must satisfy

$$\nabla f(x) + \lambda \nabla h(x) = 0,$$

for an appropriate multiplier λ. Let us pretend that we have an approximate value of λ, λ_j. The issue is how we can use this λ_j in order to find an approximation to the optimal solution x_j, and simultaneously improve the approximation of the multiplier λ_{j+1} to proceed iteratively. Obviously, the initial optimization problem is equivalent to

$$\text{Minimize}\quad f(x) + \lambda_j\ h(x)\quad \text{under}\quad h(x) = 0,$$

where we have incorporated, in a trivial manner, λ_j. To solve this problem, we introduce a quadratic penalization of the type we discussed earlier, and treat the problem of minimizing the cost function

$$f(x) + \lambda_j\ h(x) + \frac{r_j}{2}\,|h(x)|^2$$

for some penalization parameter r_j. The optimality condition for this problem is

$$\nabla f(x) + \lambda_j \nabla h(x) + r_j h(x)\nabla h(x) = 0.$$

If we assume that the approximation x_j is good, it must be close to the true optimal solution x, so therefore, if we compare the optimality conditions for both problems,

$$\nabla f(x) + \lambda \nabla h(x) = 0,$$
$$\nabla f(x_j) + \lambda_j \nabla h(x_j) + r_j h(x_j)\nabla h(x_j) = 0,$$

we come to the conclusion that

$$\lambda_j + r_j h(x_j) \approx \lambda.$$

These heuristic ideas (which can indeed be conveniently formalized) lead to the following iterative scheme:

1. **Initialization**. Take λ_1, $r_1 > 1$, $c > 1$, and tolerance $\epsilon > 0$.

2. **New approximation**. *Find an approximation x_j of the nonconstrained minimization problem for*

$$f(x) + \lambda_j \; h(x) + \frac{r_j}{2} \, |h(x)|^2 \, .$$

3. **Stopping criterion**. *If*

$$|\nabla f(x_j) + \lambda_j \; \nabla h(x_j)| < \epsilon,$$

stop: x_j is a sufficiently good approximation. Else, continue.

4. **Update**. *Update the values of the multiplier and the penalization parameters by taking*

$$\lambda_{j+1} = \lambda_j + r_j h(x_j), \quad r_{j+1} = c r_j.$$

Go to step 2.

For a general NLPP where both types of constraints in the form of equalities and inequalities are present, various tricks may reduce the problem to the case of equality constraints such as introducing new slack variables. But the following algorithm, in the spirit of the previous ideas, can also be used. The NLPP is now

$$\text{Minimize} \quad f(x)$$

subject to

$$g(x) \le 0, \quad h(x) = 0.$$

For notational convenience, let us set the augmented Lagrangian

$$L(x, \mu, \lambda, r) = f(x) + \mu \; g(x) + \lambda \; h(x) + \frac{r}{2} \left(\sum_i \max \{0, g_i(x)\}^2 + \sum_k h_k(x)^2 \right).$$

1. **Initialization**. *Choose $\mu_1 \ge 0$, λ_1, $r_1 > 0$, $c > 1$, and tolerance $\epsilon > 0$.*
2. **New approximation**. *Find an approximation x_j for the minimum of the augmented Lagrangian $L(x, \mu_j, \lambda_j, r_j)$. Recall that the functions*

$$\max \{0, g_i(x)\}^2$$

are differentiable, so that we can apply a typical algorithm for unconstrained problems such as the conjugate gradient or quasi-Newton.

3. **Stopping criterion.** *If*

$$|\nabla f(x_j) + \mu_j \, \nabla g(x_j) + \lambda_j \, h(x_j)| < \epsilon, \quad |\mu_j \, g(x_j)| < \epsilon, |h(x_j)| < \epsilon,$$

stop and take as a good approximation x_j.
4. **Update.** *Update the values of the multipliers and penalization parameter by the formulas*

$$\lambda_{j+1} = \lambda_j + r_j h(x_j), \quad \mu_{j+1} = \max\{0, \mu_j + r_j g(x_j)\}, \quad r_{j+1} = cr_j.$$

Go to step 2.

6. FINAL REMARKS

The task of searching for optimal global solutions of optimization problems is a tremendously complex and subtle job. Our readers may have the (false) impresion that the methods that have been described throughout this chapter, or others that have been omitted, are sufficient for implementing realistic problems. This is far from the truth. In reality, the gradient methods studied earlier yield, with a great deal of success, local minima of objective functionals. Graphically, we can say that by using a particular first approximation taken "ad hoc," the algorithm will lead us to the local minimum located in the "valley of atraction" of our initial guess. If we change this initial iterate, and this lies in a different valley, our final approximation will furnish a different local minimum. What is dramatic is that most likely, neither of the two will be the one global minimum we are seeking; but even if one of those were truly the global minimum, there is no way we can be certain about that. All kinds of heuristic algorithms have been developed over the years to approximate global solutions of technological problems of large dimension. Most of them incorporate a random ingredient of some sort. Among them we can cite decomposition techniques, montecarlo algorithms and their variants, simulated annealing and its variants, and genetic algorithms.

This paragraph is devoted to insisting on the importance of convexity from the horizon of the numerical approximation of optimal solutions. How important is the information that the objective function and the constraints are convex as regards the global optimal solutions? The answer is given in Theorem 3.14: A convex function defined over a convex set can have at most one valley, so that any algorithm providing approximations for a local minimum yields a global minimum as well. It is essentially the only situation in which we can guarantee that algorithms catch global minima. Otherwise, the problem of finding and approximating global optimal solutions does not have a rigorous solution.

Finally, let us stress again that many of the algorithms presented in this chapter have been implemented in commercial software packages (see the introduction to this chapter). The person who needs to solve optimization problems on a regular basis (linear and nonlinear programming) will find it quite convenient to rely on those computing tools and concentrate on modeling issues directly related to optimization problems. Furthermore, effective implementation of algorithms requires a good deal of experimentation, since the tuning of parameters is an essential ingredient for success. This does not mean that writing a program for a conjugate gradient method, for instance, in one of the typical computing languages is not a good exercise. Indeed, once students have had such an experience they will really appreciate the importance of the work done by specialists in numerical optimization. Some practice is proposed in the exercises that follow.

7. EXERCISES

1. By using one of the typical line search methods, find the minima for the following functions, starting out at the given initial points.
   1. $f(x) = x^4 - x^2 + x - 1$, $x_0 = 1$;
   2. $g(x) = x^{16} + 3x^{14} - x^7 - 3$, $x_0 = -1$;
   3.
$$h(x) = \int_{-1}^{x} \frac{se^s + s - 1}{2e^2 + 3} \, ds, \qquad x_0 = -1.$$

2. Argue that by minimizing the primitive
$$F(x) = \int_0^x f(s) \, ds$$
   we can find some solutions of the (nonlinear) equation $f(x) = 0$. Apply this to looking for some solutions of the equations
$$x^7 + x^4 + 1 = 0, \quad \log x + 6x = 0.$$

3. Consider the function

$$f(x) = \frac{1}{100} \left(x^6 - 30x^4 + 192x^2 + 7x^3 \right).$$

We seek to find the global minimum of f over the real line **R**. Apply one of the line search methods starting with

1. $x_1 = 1$;
2. $x_1 = -1.2$;
3. $x_1 = 5$;
4. $x_1 = 3$;
5. $x_1 = -8$;
6. $x_1 = -3$.

What are your conclusions? Are you sure about what the global minimum of f is? Can you think of any way to be certain about this global minimum in this case?

4. Try to approximate different minima for the functions

$$f(x) = \frac{2 - \sin |x| + x^2}{2 + \sin x}, \quad g(x) = \sin \left(\frac{x^2}{2} \right) + \sqrt{|x + 1|},$$

by using one of the line search algorithms starting with different initial guesses.

5. The linear system $Ax = b$ can be numerically solved by minimizing the function

$$f(x) = \frac{1}{2} |Ax - b|^2.$$

If the system is solvable, the global minimum should vanish; if it is not, the point of minimum is, in a sense, the closest to a solution. Apply a steepest-descent algorithm to numerically solve the systems

1.

$$A = \begin{pmatrix} 3 & -1 \\ -1 & 1 \end{pmatrix}, \quad b = \begin{pmatrix} 1 \\ 2 \end{pmatrix};$$

2.

$$A = \begin{pmatrix} 3 & -1 \\ -1 & 1 \\ 2 & 1 \end{pmatrix}, \quad b = \begin{pmatrix} 1 \\ -1 \\ 1 \end{pmatrix};$$

3.

$$A = \begin{pmatrix} 3 & 1 & -1 \\ 1 & 2 & 2 \\ -1 & 2 & 1 \end{pmatrix}, \quad b = \begin{pmatrix} 0 \\ -1 \\ 1 \end{pmatrix}.$$

6. When the matrix A is symmetric and positive definite, the linear system $Ax = b$ is the equation for critical points of the associated quadratic form

$$Q(x) = \frac{1}{2}x^T A x - x^T b.$$

Argue why this is so and find an approximate solution to the linear system by minimizing the quadratic form in the case

$$A = \begin{pmatrix} 1 & 1 & -1 \\ 1 & 2 & 1/4 \\ -1 & 1/4 & 3 \end{pmatrix}, \quad b = \begin{pmatrix} 1 \\ 1 \\ 1 \end{pmatrix}.$$

7. Find the minimum of the function

$$P(x, y) = x^2 + 2y^2 - 2x - 8y$$

and approximate it by using the steepest descent starting out at $(0, 0)$.
8. Approximate the minimum of the function

$$P(x, y) = 2x^2 - 2xy + y^2 + 2x - 2y$$

by the steepest-descent method. What results do you get? Approximate the same problem by a conjugate gradient method and compare your results.
9. Same as in the previous exercise for the function

$$P(x, y) = \frac{1}{4}(x^4 - 4xy + y^4).$$

10. Examine and study the following minimization problems with the given initial points:
   1. $100(x^2 - y)^2 + (1 - x)^2$, $(-1.2, 1)$;
   2. $(x^2 - y)^2 + (1 - x)^2$, $(-2, -2)$;
   3. $(x^2 - y)^2 + 100(1 - x)^2$, $(2, -2)$;

4. $100(x^3 - y)^2 + (1 - x)^2$, $(-1.2, 1)$.

11. Approximate the minimization problem for

$$f(x_1, x_2, x_3, x_4) = (x_1 + 10x_2)^2 + 5(x_3 - x_4)^2 + (x_2 - 2x_3)^4 + (10x_1 - x_4)^4,$$

with initial guess $(3, -1, 0, 1)$.

12. We would like to approximate the minimum value of the function $f(x, y) = y - x^2$ over the unit circle $x^2 + y^2 = 1$.

1. Find such minimum in an exact fashion.

2. Approximate such a solution by using penalized functions of the type

$$f_n(x, y) = y - x^2 + n(x^2 + y^2 - 1)^n$$

for increasing values of n. Compare your results.

13. Set up a suitable strategy for approximating the optimal solution of

$$\text{Minimize} \quad \frac{1}{2} x^T A x$$

subject to

$$bx = c,$$

where A is a symmetric positive definite matrix. Apply this to solving the case for

$$A = \begin{pmatrix} 3 & 3 & 1 \\ 3 & 5 & 3 \\ 1 & 3 & 3 \end{pmatrix}, \quad b = (1 \ \ 1 \ \ 1), \quad c = 1.$$

14. Same exercise as the previous one but changing the equality restriction $bx = c$ to the inequality $bx \le c$. How would you approximate a similar problem under a quadratic restriction of the type

$$x^T B x + bx \le c$$

with B symmetric and positive semidefinite? Apply your method to solving

$$\text{Minimize} \quad |x|^2$$

under

$$(2x_1 - x_2)^2 + (x_3 - 2)^2 \le 1.$$

15. Choose some of the exercises proposed in Chapter 3 for a particular data set and approximate their solutions.

Chapter 5

Variational Problems and Dynamic Programming

1. INTRODUCTION

We start in this chapter the analysis of optimization problems of a different nature. Specifically, this chapter is devoted to variational problems of finding the infimum of the integrals

$$\int_\Omega F(x, u(x), \nabla u(x))\, dx, \tag{5–1}$$

where $\Omega \subset \mathbf{R}^N$, the functions $u : \Omega \to \mathbf{R}$ must be differentiable, and they typically are also constrained in some other way such as having their boundary values on $\partial\Omega$ fixed by some preassigned function u_0, i.e., $u = u_0$ on $\partial\Omega$. The integrand (or Lagrangian)

$$F : \Omega \times \mathbf{R} \times \mathbf{R}^N \to \mathbf{R}$$

characterizes each such problem. The proposed task consists in finding a function U, admissible according to the restrictions we have imposed on competing functions, such that the integral

$$\int_{\Omega} F(x, U(x), \nabla U(x)) \, dx$$

is smaller than (or equal to) the same integral for any other feasible function u. If we use the notation

$$I(u) = \int_{\Omega} F(x, u(x), \nabla u(x)) \, dx,$$

we are interested in understanding the optimization problem formulated as

$$\text{Minimize} \quad I(u)$$

subject to further constraints on the functions u, such as for instance,

$$u(x) = u_0(x), \quad x \in \partial\Omega.$$

It is therefore an optimization problem in which admissible functions replace feasible vectors.

We will be mainly concerned, in this introductory chapter for variational problems, about the one-dimensional situation where $\Omega = (a, b)$ is an interval in $\mathbf{R}$, and admissible functions $u : (a, b) \to \mathbf{R}$ will often be required to satisfy the boundary conditions

$$u(a) = A, \quad u(b) = B,$$

for known values A, B. In this case

$$I(u) = \int_{a}^{b} F(x, u(x), u'(x)) \, dx,$$

where now the integrand F is a function of three (or fewer) variables. We have already mentioned some of these examples in Chapter 1, and have tried to convince the reader that learning to solve this sort of problem (or at least to approximate optimal solutions appropriately) might be important. In this chapter we will study and solve more examples, and learn the main techniques in dealing with such problems.

There is a great variety of variational problems. The common ingredient is that costs are typically represented by an integral of the type indicated above. But additional constraints may vary from example to example. We can classify these constraints as follows:

1. **Boundary conditions.** One of the most common situations corresponds to having prescribed boundary values along the complete boundary $\partial\Omega$; but other possibilities include having this prescription in part of the boundary (in particular no condition on the boundary at all) and having prescribed not the values of the functions but those of the derivative or some of the derivatives.

2. **Integral constraints**, requiring of competing functions to comply with restrictions of the type

$$\int_\Omega G(x, u(x), \nabla u(x))\, dx = \alpha,$$

where

$$G : \Omega \times \mathbf{R} \times \mathbf{R}^N \to \mathbf{R}^d, \quad \alpha \in \mathbf{R}^d,$$

are known; some of these constraints could come in the form of inequalities.

3. **Pointwise constraints**, establishing that feasible functions must respect the condition

$$G(x, u(x), \nabla u(x)) = 0,$$

for all x in Ω, where again G is a known function as above, and we could also have some inequalities.

Finally, it is important to point out that some of the techniques to be discussed can be extended without much change to situations in which cost functionals include a dependence on higher derivatives (or no derivatives at all). We will also deal with some of these situations.

Another aspect, intimately connected with variational problems, refers to dynamic programming. In the discrete case, we will briefly discuss the main underlying principle leading to optimal solutions. In the continuous case, we will heuristically establish Bellman's equation of dynamic programming, which will help us to derive Pontryagin's maximum principle in the next chapter.

It is fair to stress the importance of variational problems in many fields of science and engineering. We can hardly mention all of them here: mechanics, elasticity (linear and nonlinear), continuous media, dynamics, material behavior, solid structures, fluids, etc. Some of our examples will illustrate in a simple and direct way the relevance and the role played by variational formulations and techniques. This is not surprising, since nature, as well as human beings, is, in one way or another always looking for the best.

2. THE EULER–LAGRANGE EQUATION: EXAMPLES

The Euler–Lagrange equation (E-L) associated with a variational problem plays the same role as the necessary conditions of optimality in a programming problem (KKT conditions). As we now know, such necessary conditions of optimality furnish restrictions that optimal solutions (possibly among other feasible vectors) must satisfy. By exploiting such conditions, optimal solutions can be found or approximated explicitly in a variety of situations. We have also stressed the central role played by the notion of convexity in ensuring that optimality conditions are in fact sufficient for optimal solutions. It is therefore not surprising that convexity will also be central to this and the next chapters: Convexity is always the key concept in minimization problems of any type.

In general, a variational problem is characterized by an integral cost functional

$$I(u) = \int_\Omega F(x, u(x), \nabla u(x)) \, dx,$$

where $\Omega \subset \mathbf{R}^N$, $u : \Omega \to \mathbf{R}$ are differentiable, and the cost integrand

$$F(x, \lambda, \xi) : \Omega \times \mathbf{R} \times \mathbf{R}^N \to \mathbf{R}$$

is assumed to be differentiable, in fact twice differentiable, with respect to the variables (λ, ξ). Note that

$$\xi = (\xi_1, \xi_2, \ldots, \xi_N).$$

We will first focus on the situation in which we have additional feasibility constraints of the type

$$u(x) = u_0(x), \quad x \in \partial\Omega,$$

since this is the most typical situation. Later, when restricting to dimension $N = 1$, we will consider other cases.

The next result is the clue to finding optimal solutions for this sort of problem.

Theorem 5.1   (Euler–Lagrange equation) Under the setting described above:
1. If u is an optimal solution, then u must also be a solution of the problem (E-L)

$$\text{div}\left(F_\xi(x, u(x), \nabla u(x))\right) = F_\lambda(x, u(x), \nabla u(x)) \quad \text{in} \quad \Omega,$$
$$u = u_0 \quad \text{on} \quad \partial\Omega.$$

2. If u satisfies E-L and F is convex with respect to the variables (λ, ξ) for each fixed $x \in \Omega$, then u is also an optimal solution of the variational problem.
3. If in addition, F is strictly convex with respect to (λ, ξ) for each $x \in \Omega$, the optimal solution u, if it exists, is unique.

Before explaining where this E-L equation comes from, it might be important to become convinced of its relevance and applicability in solving some variational problems. We are going to look at some particular situations including some of the cases discussed in the first chapter. Most of these examples correspond to the one-dimensional case, when $N = 1$ and Ω is in fact an open interval (a, b) on the real line. The E-L equation is now a second order ordinary differential equation completed with the appropriate boundary values

$$\frac{d}{dx}\left[F_\xi(x, u(x), u'(x))\right] = F_\lambda(x, u(x), u'(x)), \quad x \in (a, b),$$
$$u(a) = A, \quad u(b) = B,$$

where A and B are typically given. In this one-dimensional situation, we will cover several possibilities so as to gain familiarity with the E-L equation.

1. Assume, to begin with, the most simple situation, in which F depends on ξ exclusively. In this case, $F = F(\xi)$, and E-L simplifies to

$$\frac{d}{dx}\left[F'(u'(x))\right] = 0,$$

which in turn holds if

$$F'(u'(x)) = k,$$

a constant. Evidently, this last requirement is fulfilled if we take u' constant throughout the interval (a, b), and this is so if u is indeed the straight line joining the points (a, A), (b, B). Such a linear (affine) function is always a solution of E-L if F depends only on the derivative variable ξ. If in addition, F is convex, that linear function will be a minimizer. If even further, F is strictly convex, this linear function is the only minimizer of the problem. In case F is not convex, even though the linear function is a solution of E-L, it may not be a minimizer, as the following example shows.

Example 5.2   *(Nonconvex example) Let us take*

$$F(\xi) = e^{-\xi^2}, \quad a = A = B = 0, \quad b = 1.$$

In this situation the linear function through the points $(0,0)$, $(1,0)$ is the function u_0 vanishing identically on the interval $(0,1)$. Its cost is 1. However, we claim that the infimum of the integrals

$$I(u) = \int_0^1 e^{-u'(x)^2} \, dx$$

subject to

$$u(0) = u(1) = 0$$

vanishes. Indeed, consider the sequence of feasible functions

$$u_j(x) = j \left(x - \frac{1}{2} \right)^2 - \frac{j}{4}.$$

It is not hard to check that $I(u_j) \searrow 0$, so that the above infimum does truly vanish. Hence, u_0 is a solution of the associated E-L, but it is not a minimizer. What fails in this situation is the convexity of F. It is even true that there is no minimizer for this problem, because such a function v would have to satisfy

$$\int_0^1 e^{-v'(x)^2} \, dx = 0,$$

and this is impossible.

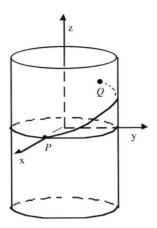

Figure 5.1. Geodesics in a cylinder.

Example 5.3 *(Geodesics in a cylinder) Let C be the cylinder of equation*

$$x^2 + y^2 = 1,$$

and let P and Q be two distinct points on C. We would like to find the shortest path over C going from P to Q. Without loss of generality we may assume that $P = (1, 0, 0)$. We will naturally pose the problem in cylindrical coordinates (r, θ, z) with

$$r = 1, \quad -\pi \leq \theta \leq \pi,$$

defining C. Let $Q = (1, \theta_0, z_0)$ (see Figure 5.1).

By symmetry considerations, it is enough to treat the case $0 < \theta_0 \leq \pi$ (what is the geodesic when $\theta_0 = 0$?). We can represent an arbitrary curve joining $(1, 0, 0)$ and $(1, \theta_0, z_0)$ in the form

$$\sigma(\theta) = (\cos \theta, \sin \theta, z(\theta)), \quad \theta \in (0, \theta_0),$$

since geodesics will necessarily meet each vertical line in C at most once (why?), and so the first two components of σ can be assumed to be of trigonometric form. In this fashion any such curve is fully determined by the function $z(\theta)$. We must also ask for

$$z(0) = 0, \quad z(\theta_0) = z_0.$$

The cost functional we must minimize is that representing the length of σ:

$$I(z) = \int_0^{\theta_0} \sqrt{1 + z'(\theta)^2}\, d\theta.$$

We know that the function

$$F(\xi) = \sqrt{1 + \xi^2}$$

is stricly convex (Chapter 3), and it depends only on the derivative variable. According to our previous discussion, the linear function

$$z(\theta) = \frac{z_0}{\theta_0}\theta$$

represents the only geodesic joining those two points. This linear function in cylindrical coordinates is an arc of an helix over C.

2. When the integrand F depends on both x and ξ, E-L becomes

$$\frac{d}{dx}\,[F_\xi(x, u'(x))] = 0,$$

or equivalently,

$$F_\xi(x, u'(x)) = \text{constant}.$$

Depending on the particular form of F, this last equation will be solvable analytically or not.

Example 5.4   *(Weierstrass's example)* Let

$$F(x, \xi) = x\xi^2, \quad x \in (0, 1),$$

and $u(0) = 1$, $u(1) = 0$ at the endpoints of the interval. In this particular example, E-L becomes, for arbitrary constants c and d,

$$xu'(x) = c, \qquad u(x) = c\log x + d.$$

Curiously enough, this family of solutions is unable to match the boundary condition at the right endpoint $u(1) = 0$, and hence the variational problem might not have optimal solutions. This is indeed the case, since the family of functions

$$u_j(x) = \begin{cases} 1, & x \in (0, 1/j), \\ -\log x / \log j, & x \in (1/j, 1), \end{cases}$$

is minimizing for

$$I(u) = \int_0^1 x u'(x)^2 \, dx, \quad u(0) = 1, \quad u(1) = 0,$$

in the sense that $I(u_j) \searrow 0$. However, for any given function u, we have $I(u) > 0$, and therefore there is no optimal solution for this variational problem.

Example 5.5 (The brachistochrone) One of the most celebrated variational problems of all time is the brachistochrone. Let us place the x-axis along the vertical direction where gravity acts, and the y-axis perpendicular to it. Assume that we have two points P and Q at different heights. Without loss of generality we may take P at the origin and $Q = (a, A)$ with both a, A positive. The task consists in determining the path joining Q and P so that a unit mass spends the least time possible in going from Q to P under the action of gravity without friction.

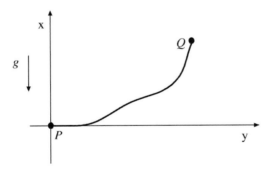

Figure 5.2. The brachistochrone.

Evidently, the optimal path can be represented in the form $y(x)$ for some function y to be determined. What we mean is that paths that are not monotone (with bumps) will obviously provide greater times than those that are monotone. Assume that $y(x)$ is one such path, so that $y(0) = 0$, $y(a) = A$.

The transit time for such a path will be given by the integral

$$\int_0^a \frac{ds}{v},$$

where ds is the differential element of arc length given by

$$ds = \sqrt{1 + y'(x)^2}\, dx,$$

and v is the velocity due to gravity at height x. According to a well known formula,

$$v = \sqrt{2gx}.$$

Altogether we are interested in finding the curve $y(x)$, $0 \le x \le a$, that minimizes the transit time integral (neglecting positive constants that do not interfere with minimization)

$$I(y) = \int_0^a \sqrt{\frac{1 + y'(x)^2}{x}}\, dx,$$

and $y(0) = 0$, $y(a) = A$. We caution the reader that the solution is neither a straight line nor a circle. In fact, in the particular case $a = A = 1$, could our readers decide which curve furnishes less transit time, the line $y = x$ or the circle $y = 1 - \sqrt{1 - x^2}$? In our problem the integrand function F is

$$F(x, \xi) = \frac{\sqrt{1 + \xi^2}}{\sqrt{x}}.$$

It is a (strictly) convex function of ξ, so that, overlooking the difficulty when $x = 0$, optimal solutions can be sought by examining the associated E-L equation. In this case we must solve

$$\frac{y'}{\sqrt{x}\sqrt{1 + (y')^2}} = \frac{1}{c}, \qquad \frac{(y')^2}{1 + (y')^2} = \frac{x}{c^2}.$$

This leads to

$$y'(x)^2 = \frac{x}{c^2 - x}, \qquad y(x) = \int_0^x \sqrt{\frac{s}{c^2 - s}}\, ds,$$

where the constant c is to be determined in such a way that

$$A = \int_0^a \sqrt{\frac{s}{c^2 - s}}\, ds.$$

Can the reader argue why we have chosen the positive sign in the above square root?

In order to find a more explicit form of the solution, we will use the change of variables in the integral for y given by

$$s(r) = \frac{c^2}{2}(1 - \cos r) = c^2 \sin^2(r/2).$$

Then

$$y(t) = c^2 \int_0^t \sin^2(r/2)\, dr = \frac{c^2}{2}(t - \sin t),$$

where

$$x(t) = \frac{c^2}{2}(1 - \cos t) = c^2 \sin^2(t/2).$$

In parametric form,

$$(x(t), y(t)) = (C(1 - \cos t), C(t - \sin t)), \quad 0 \le t \le t_0,$$

is the solution. It already satisfies $x(0) = y(0) = 0$. The constants C and t_0 must be found by imposing $x(t_0) = a$, $y(t_0) = A$. This curve is an arc of a cycloid (Figure 5.3).

Figure 5.3. The cycloid.

3. We finally analyze the case in which $F = F(\lambda, \xi)$. In this situation the E-L equation has the form

$$\frac{d}{dx} \left[F_\xi(u(x), u'(x)) \right] = F_\lambda(u(x), u'(x)).$$

It is a matter of careful arithmetic to see that this equation (in the case in which $F = F(\lambda, \xi)$ and only in this case) can be rewritten

$$\frac{d}{dx} \left[F(u(x), u'(x)) - u'(x) F_\xi(u(x), u'(x)) \right] = 0,$$

and this leads to

$$F(u(x), u'(x)) - u'(x) F_\xi(u(x), u'(x)) = \text{ constant.}$$

In some situations this form of the equation can be more appropriate when one is looking for solutions. But on other occasions it may not be so.

Example 5.6   (*Minimal surfaces of revolution*) *We would like to identify functions $u(x)$ defined on the interval (a, b) such that $u(a) = A$, $u(b) = B$, and whose graphs generate, by revolution around the X-axis, the surface with least area (Figure 5.4).*

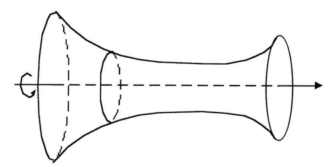

Figure 5.4. A surface of revolution.

We know from calculus that this area is given, except for a positive multiplicative constant, by the integral

$$I(u) = \int_a^b u(x) \sqrt{1 + u'(x)^2} \, dx.$$

We seek the function that minimizes this integral among all those satisfying the conditions at both endpoints. The integrand for this example is

$$F(\lambda, \xi) = \lambda\sqrt{1 + \xi^2}.$$

This function is convex in ξ for fixed λ provided $\lambda \geq 0$. It is even strictly convex in ξ if $\lambda > 0$. However, it is not jointly convex in (λ, ξ) (why?). Hence in principle, we cannot apply Theorem 5.1. However, it is true that convexity is needed only with respect to ξ for fixed (x, λ) for the conclusion of this result to be valid. This fact is beyond the scope of this text, but it is important to have this more general fact in mind for treating this example.

Theorem 5.7   ([19]) If u is the unique solution of E-L and F is convex with respect to the variable ξ for each fixed $x \in \Omega$ and $\lambda \in \mathbf{R}$, then u is also an optimal solution of the variational problem.

We can therefore proceed with the study of the E-L equation to look for optimal solutions.

The second form of the E-L equation for this example is

$$u\sqrt{1 + (u')^2} - u'\frac{uu'}{\sqrt{1 + (u')^2}} = c.$$

After several manipulations, separating variables, we arrive at

$$\frac{du}{\sqrt{u^2 - c^2}} = \frac{dx}{c}.$$

In this particular case taking the positive or negative sign for the square root is irrelevant. We invite our readers to check this. A change of variables involving hyperbolic trigonometric functions leads us to the final form of the solution

$$u(x) = c\,\cosh\left(\frac{x}{c} + d\right),$$

which is a catenary curve. This is the unique solution.

When we come to adjust the values at the endpoints, we find some difficulties. Assume, for the sake of simplicity, that we demand $u(0) = 1$, $u(b) = B$. The first condition allows us to put

$$c = \frac{1}{\cosh d},$$

while the second implies

$$B = \frac{\cosh(b \cosh d + d)}{\cosh d}.$$

This condition cannot always be satisfied. Indeed, if we keep in mind that

$$\cosh t > |t|,$$

for any real t, it turns out that

$$B > \frac{b \cosh d + d}{\cosh d} \geq \frac{b \cosh d - |d|}{\cosh d} = b - \frac{|d|}{\cosh d} \geq b - 1.$$

This chain of inequalities means that if at the outset we have $B \leq b - 1$, there is no way we can adjust the value at the right endpoint, and consequently, the problem may not have optimal solutions.

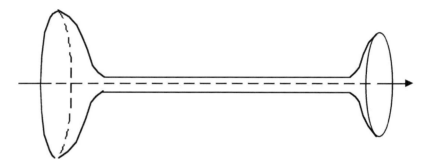

Figure 5.5. Toward a limit situation.

This observation can also be illustrated physically in a rather appealing way. The form adopted by a soap film that adheres to a (non planar) ring is related to the surface tension in such a way that the shape adopted by the soap film will be the one minimizing this surface tension. This quantity is proportional to the area generated by the surface, so that determining the optimal shape is equivalent to finding the minimal surface area. In the case of surfaces of revolution, such as those we are discussing in this example, we can imagine two concentric rings of different but close radii and a certain small distance apart. A soap film glued to the two rings will exhibit the form given by the catenary we have found earlier. But as we start moving the two rings farther away from each other, the film will stretch until a certain point where the film will degenerate into two circles around each ring. This transition is exactly reflected by our computations.

The explanation of this fact relies on an understanding of our variational problem. When the two rings are close together, the catenary provides the minimal surface of revolution with area smaller than that of the two circles around the rings. But as we move the rings apart, the area of the catenary grows to a point where it equals that of the two circles. If we push further, no surface of revolution will have a smaller area than the two circles, and hence the soap film collapses. This optimal solution (the two disks) cannot be described by a surface of revolution generated by the graph of a function u, so that in reality our variational problem does not have an optimal solution. A more general formulation of the problem would be required to incorporate these special solutions. However, it is true that we can approximate this special solution as much as we desire by a sequence of admissible surfaces of revolution u_j according to Figure 5.5.

4. We finally examine various easy situations in higher dimensions.

Example 5.8 (Dirichlet's integral) Let $u : \Omega \subset \mathbf{R}^N \to \mathbf{R}$ be a function satisfying $u = u_0$ over $\partial\Omega$, where u_0 is a fixed function. We would like to identify the function u minimizing the integral

$$I(u) = \frac{1}{2} \int_\Omega |\nabla u(x)|^2 \, dx.$$

We can imagine the situations in which $N = 2$ or $N = 3$, and Ω is a circle in the plane or a ball in space. It is therefore a problem of the calculus of variations. We notice that the integrand

$$F(\xi) = \frac{1}{2} |\xi|^2$$

is strictly convex, so that there exists an optimal solution, and it is unique (see Theorem 5.1.) Such a function must be a solution of the E-L equation. In this case, the equation is

$$\text{div}\,(F_\xi(\nabla u(x))) = 0,$$

i.e.,
$$\Delta u = 0.$$

Thus, the function minimizing the square of the norm of the gradient is the harmonic function respecting the given boundary values on $\partial \Omega$. This is often interpreted by saying that the harmonic function corresponds to a stable equilibrium state with respect to an energy proportional to the square of the gradient.

Example 5.9   (Wave equation) For the particular case in which
$$F(\xi_1, \xi_2) = \frac{1}{2} \left(\xi_1^2 - \xi_2^2 \right),$$

the E-L equation is precisely the wave equation
$$u_{xx} - u_{yy} = 0.$$

In this case, however, F is not convex, so that we cannot talk about minimization. Nevertheless, for this type of equation there is a rich theory of variational principles in mechanics where Hamilton's principle of minimum action is postulated and the central role is moved from the concept of minimizer to that of stationary state.

A simple situation may help us in understanding a bit better what we mean by the previous sentences. Suppose a particle of mass m travels in a straight line under the action of a force field $F(t, x)$ depending on position and time. If $u(t)$ indicates the position of the particle at time t, Newton's law ensures that
$$\frac{d}{dt}(mu'(t)) = F(t, u(t)).$$

The question is, can we invent a function $L(t, \lambda, \xi)$ such that the associated E-L equation for the functional
$$I(u) = \int_0^{t_0} L(t, u(t), u'(t)) \, dt$$

turns out to be exactly Newton's law? In this simplified situation this is not difficult. If we assume that the vector field F is conservative with potential $U(t, \lambda)$ such that $U_\lambda = -F$, then we can take
$$L(t, \lambda, \xi) = \frac{1}{2} \xi^2 - U(t, \lambda).$$

This function is convex in ξ. By Theorem 5.7, we would reach the conclusion that the movement of the particle would take place according to a principle of least energy (least action) as measured by I. The function L is called the Lagrangian, and the functional I is called the action integral. A full study of these topics greatly exceeds the scope of this book.

Example 5.10 (Minimal surfaces) We consider the problem of minimal surfaces, not necessarily of revolution around an axis. Given a plane region Ω and a function u_0 over $\partial\Omega$ (the ring), the area of the graph of a function u defined in Ω is given by the integral

$$I(u) = \int_\Omega \sqrt{1 + |\nabla u(x)|^2}\, dx.$$

We are looking for a function u minimizing this integral among all those functions having the same values as u_0 over $\partial\Omega$. Since the function

$$F(\xi) = \sqrt{1 + |\xi|^2}$$

is strictly convex, if there is a solution it must be unique. The E-L equation is

$$\operatorname{div}\left(\frac{\nabla u}{\sqrt{1 + |\nabla u|^2}}\right) = 0.$$

This partial differential equation is a complicated equation for a number of reasons far beyond the scope of this text. In the case $N = 2$, the equation may be rewritten as

$$(1 + u_y^2)u_{xx} - 2u_x u_y u_{xy} + (1 + u_x^2)u_{yy} = 0.$$

3. THE EULER–LAGRANGE EQUATION: JUSTIFICATION

After being convinced through numerous examples of the importance of the E-L equation in finding optimal solutions for variational problems, it is worthwhile to explore where this equation comes from and why optimal solutions must also be solutions of this equation. We will dwell on these issues at the level of the underlying ideas and skip several technical issues that are irrelevant to our discussion. We will first treat the one-dimensional situation in order to better understand its genesis, and then we will indicate the changes for the higherdimensional case.

We remind the reader that we would like to justify the following important result.

Theorem 5.1   (Euler–Lagrange equation) Under the setting described earlier:

1. If u is an optimal solution, then u must also be a solution of the problem (E-L)

$$\operatorname{div}\left(F_\xi(x, u(x), \nabla u(x))\right) = F_\lambda(x, u(x), \nabla u(x)) \quad \text{in} \quad \Omega,$$
$$u = u_0 \quad \text{on} \quad \partial\Omega.$$

2. If u satisfies E-L and F is convex with respect to the variables (λ, ξ) for each fixed $x \in \Omega$, then u is also an optimal solution of the variational problem.
3. If in addition, F is strictly convex with respect to (λ, ξ) for each $x \in \Omega$, the optimal solution u, if it exists, is unique.

For the one-dimensional situation the equation simplifies to

$$\frac{d}{dx}\left[F_\xi(x, u(x), u'(x))\right] = F_\lambda(x, u(x), u'(x)),$$

together with boundary conditions $u(a) = A$, $u(b) = B$, where

$$F = F(x, \lambda, \xi), \quad I(u) = \int_a^b F(x, u(x), u'(x))\, dx.$$

Let φ be a fixed function satisfying the requirement $\varphi(a) = \varphi(b) = 0$, and let us consider the function of a single variable

$$g(t) = I(u + t\varphi) = \int_a^b F(x, u(x) + t\varphi(x), u'(x) + t\varphi'(x))\, dx,$$

where we are assuming that u is an optimal solution yielding the least value of the above integrals among all feasible functions. For each choice of $t \in \mathbf{R}$, the function $u + t\varphi$ turns out to be admissible, because φ vanishes on the endpoints of the interval (a, b). Therefore, g has a (global) minimum for $t = 0$. A necessary condition for the occurrence of such a minimum is that the derivative vanish at such a point. If we differentiate under the integral sign in the definition of g and evaluate at $t = 0$, we obtain

$$0 = \int_a^b [F_\lambda(x, u(x), u'(x))\varphi(x) + F_\xi(x, u(x), u'(x))\varphi'(x)] \, dx.$$

Integrating by parts in the second term and bearing in mind that $\varphi(a) = \varphi(b) = 0$, we have

$$0 = \int_a^b \left[F_\lambda(x, u(x), u'(x)) - \frac{d}{dx} F_\xi(x, u(x), u'(x)) \right] \varphi(x) \, dx.$$

Since φ is arbitrary, save for its vanishing values at both endpoints, the previous identity can happen only if the expression within brackets identically vanishes: the E-L equation. This is the first part of the theorem.

Let us now suppose that the integrand $F(x, \lambda, \xi)$ is jointly convex in the variables (λ, ξ) for each fixed $x \in (a, b)$, and that u is a solution of E-L together with the appropriate boundary conditions. Let v be any other feasible function such that $v(a) = A$, $v(b) = B$. By the convexity of F (Chapter 3),

$$
\begin{aligned}
I(v) - I(u) &= \int_a^b [F(x, v(x), v'(x)) - F(x, u(x), u'(x))] \, dx \\
&\geq \int_a^b [F_\lambda(x, u(x), u'(x))(v(x) - u(x)) \\
&\quad + F_\xi(x, u(x), u'(x))(v'(x) - u'(x))] \, dx.
\end{aligned}
$$

If we integrate by parts in the second term as before, we notice that we obtain exactly the E-L equation for u, so that if it is indeed a solution, the conclusion is

$$I(v) - I(u) \geq 0,$$

and u is truly an optimal solution for the problem. This is the sufficiency part of the theorem.

Finally, we would like to prove the uniqueness of optimal solutions under the strict convexity of F. The easiest way of dealing with the strict convexity in this context consists in requiring that the equality

$$f\left(\frac{1}{2}x + \frac{1}{2}y\right) = \frac{1}{2}f(x) + \frac{1}{2}f(y)$$

automatically imply $x = y$ if f is a strictly convex function. Let us try to reach one such situation when we assume that $F(x, \cdot, \cdot)$ is strictly convex.

Imagine that our variational problem admits two optimal solutions u, v. Due to the convexity, it is not hard to deduce

$$\frac{1}{2}I(u) + \frac{1}{2}I(v) - I\left(\frac{1}{2}u + \frac{1}{2}v\right) \geq 0.$$

If we denote by m the value of the minimum, i.e., $I(u) = I(v) = m$, then

$$m - I\left(\frac{1}{2}u + \frac{1}{2}v\right) \geq 0.$$

But on the other hand, since m is the value of the minimum,

$$I\left(\frac{1}{2}u + \frac{1}{2}v\right) \geq m.$$

We can conclude that in fact,

$$I\left(\frac{1}{2}u + \frac{1}{2}v\right) = m,$$

and hence

$$0 = \int_a^b \left[\frac{1}{2}F(x, u(x), u'(x)) + \frac{1}{2}F(x, v(x), v'(x)) \right.$$
$$\left. - F\left(x, \frac{1}{2}u(x) + \frac{1}{2}v(x), \frac{1}{2}u'(x) + \frac{1}{2}v'(x)\right)\right]\, dx.$$

But the previous integrand is nonnegative, again by the convexity of F. The only possibility for a nonnegative function whose integral vanishes is to vanish identically, so that

$$\frac{1}{2}F(x, u(x), u'(x)) + \frac{1}{2}F(x, v(x), v'(x))$$
$$- F\left(x, \frac{1}{2}u(x) + \frac{1}{2}v(x), \frac{1}{2}u'(x) + \frac{1}{2}v'(x)\right) \equiv 0.$$

By the remark made earlier, this implies that $u = v$, and the optimal solution is therefore unique if it exists.

For the case of a problem in several variables, the argument is formally the same. The changes relate to the way in which integration by parts must be performed through the divergence theorem. For instance, if φ vanishes on $\partial\Omega$, so that contributions from the boundary drop, we will have

$$0 = \int_{\Omega} \left[F_\lambda(x, u(x), \nabla u(x))\varphi(x) + F_\xi(x, u(x), \nabla u(x))\nabla\varphi(x)\right] dx$$
$$= \int_{\Omega} \left[F_\lambda(x, u(x), \nabla u(x)) - \operatorname{div}\left(F_\xi(x, u(x), \nabla u(x))\right)\right] \varphi(x) \, dx.$$

The proof proceeds accordingly.

4. NATURAL BOUNDARY CONDITIONS

It is interesting to stress the way in which we have found E-L. To emphasize what we mean, we are going to deal with a typical situation in which the value at one of the two endpoints is free in a variational problem in dimension one. We would like to find the minimum of the integral

$$I(u) = \int_a^b F(x, u(x), u'(x)) \, dx,$$

where competing functions u are required to satisfy $u(a) = A$ exclusively, but nothing is demanded at the right endpoint, so that the set of feasible functions is larger compared to the situation in which we fix the value at that endpoint. We suspect that E-L together with $u(a) = A$ might be insufficient to completely determine the optimal solution u. Somehow, the condition of having a free endpoint must impose a further condition on optimal solutions. This is indeed so. If we return to the derivation of the E-L equation, we observe that the stage at which the conditions at the endpoints were important was in the choice of the auxiliary function φ, a function that must vanish at both a and b. If now we must leave the value at b free, this amounts to considering φ arbitrary except for $\varphi(a) = 0$, but nothing is required at b. This information was used at the point of the integration by parts. If we do not have $\varphi(b) = 0$, we would get

$$0 = \int_a^b \left[F_\lambda(x, u(x), u'(x)) - \frac{d}{dx} F_\xi(x, u(x), u'(x)) \right] dx$$
$$+ F_\xi(b, u(b), u'(b))\varphi(b).$$

If first we restrict attention to all φ's vanishing at b (because those φ's are also eligible), we would conclude, as before, that the E-L equation should hold. But once we have this information at our disposal, the above identity forces us to have

$$F_\xi(b, u(b), u'(b))\varphi(b) = 0.$$

Since $\varphi(b)$ can be chosen arbitrarily, this implies

$$F_\xi(b, u(b), u'(b)) = 0,$$

which is the so-called transversality or natural boundary condition at b. This is an additional condition that optimal solutions to the variational problem with the right endpoint free must satisfy. The same observations apply to the left endpoint.

Example 5.11   *Let us try to find the least value that the integrals*

$$I(u) = \frac{1}{2} \int_0^{\log 2} \left[(u'(x) - 1)^2 + u(x)^2 \right] dx$$

can take on among all functions u. This is a situation in which both endpoints are free, so that since the integrand

$$F(x, \lambda, \xi) = (\xi - 1)^2 + \lambda^2$$

is strictly convex, the optimal solution is found by solving the problem

$$u''(x) - u(x) = 0, \quad u'(0) = u'(\log 2) = 1.$$

The unique solution is

$$u(x) = \frac{1}{3}e^x - \frac{2}{3}e^{-x}.$$

Another possibility occurs when the values at the endpoints are restricted by the inequalities

$$B_1 \le u(b) \le B_2.$$

In this case we proceed as follows. First, we examine the transversality condition

$$F_{\xi}(b, u(b), u'(b)) = 0.$$

If this determines the optimal solution u so that it is feasible, then this is our optimal solution. If it is not so because the value $u(b)$ does not lie in the interval $[B_1, B_2]$, then the optimal solution will have either $u(b) = B_1$ or $u(b) = B_2$, depending on whether $u(b) < B_1$ or $u(b) > B_2$, respectively. This rule would require further comments and a full discussion on convexity, but we will take it as valid, and indeed it is correct in many regular cases. We will come back to this issue later.

Example 5.12   Consider the following easy situation:

$$Minimize \quad \int_0^{\log 2} \left[u'(x)^2 + (u(x) - 2)^2 \right] dx$$

subject to

$$2 \le u(0) \le 3, \quad u(\log 2) = 1.$$

The E-L equation together with endpoint conditions reads

$$u''(x) = u(x) - 2, \qquad u'(0) = 0, \quad u(\log 2) = 1.$$

The solution is

$$u(x) = 2 - \frac{2}{5}\left(e^x + e^{-x} \right).$$

We notice that $u(0) = 6/5$, so that this solution is not admissible for our optimization problem. Since, however, the value $u(0)$ is smaller than the permissible values at 0, we conclude that the optimal solution will be the solution of the problem

$$u''(x) = u(x) - 2, \qquad u(0) = 2, \quad u(\log 2) = 1.$$

The optimal solution is thus

$$u(x) = -\frac{2}{3}\left(e^x - e^{-x}\right) + 2.$$

5. VARIATIONAL PROBLEMS UNDER INTEGRAL AND POINTWISE RESTRICTIONS

In this section we would like to consider variational problems in which in addition to having constraints on the values on both endpoints, we must respect conditions expressed in terms of equalities and/or inequalities of the type

$$\int_a^b G(x, u(x), u'(x))\,dx \le \alpha, \quad \int_a^b H(x, u(x), u'(x))\,dx = \beta.$$

Notice that both G and H could be vector-valued, so that in reality we may have several integral constraints. The vectors α and β are given. In this situation we are willing to accept as competing functions those respecting all these integral constraints, and among them we would like to find the one(s) realizing the least value of the integrals

$$\int_a^b F(x, u(x), u'(x))\,dx.$$

As remarked, we might also have constraints at the endpoints. As a matter of fact, notice that

$$u(a) = A, \quad u(b) = B$$

is equivalent to

$$u(a) = A, \quad \int_a^b u'(x)\,dx = B - A,$$

and we may incorporate the function $G_0(x, \lambda, \xi) = \xi$ as one integral constraint. We will understand this condition in this way throughout this section.

   As might be expected from the experience we already have in mathematical programming, we have to consider multipliers associated with all the integral constraints, one for each such restriction. Thus we will have to work with the augmented integrand

$$\tilde{F}(x, \lambda, \xi) = F(x, \lambda, \xi) + yG(x, \lambda, \xi) + zH(x, \lambda, \xi).$$

The practical process of finding such optimal solutions is slightly different from that of mathematical programming, although it is based on the same underlying ideas. Let us first treat the case of equality constraints, so that the function G is not present.

Proposition 5.13   *Assume that there is a vector of numbers z (the vector of multipliers) such that the auxiliary integrand*

$$\tilde{F} = F + zH$$

turns out to be convex in (λ, ξ) (in fact, only the convexity with respect to ξ is needed) for each fixed $x \in (a, b)$. If u is the unique solution of the E-L problem associated with $\tilde{F}$,

$$\frac{d}{dx}\left[\tilde{F}_\xi(x, u(x), u'(x))\right] = \tilde{F}_\lambda(x, u(x), u'(x)),$$

with $u(a) = A$, then u is an optimal solution of the problem under the integral constraints corresponding to the vector β determined by u itself

$$\beta = \int_a^b H(x, u(x), u'(x))\, dx.$$

If the convexity of $\tilde{F}$ is strict, then the uniqueness of the optimal solution follows. The proof of this result reduces to applying the convexity part of Theorem 5.1 to $\tilde{F}$. It is left to the interested reader.

The way in which this result is used in practical computations consists of two steps. First, the E-L equation for $\tilde{F}$ is solved incorporating the multipliers z in the whole process as parameters, so that we obtain a whole family of solutions, one for each z. Afterwards, these multipliers are adjusted in such a way that the corresponding optimal solution yields the appropriate value for the integral constraint.

Example 5.14   *Find the function u minimizing the integral of the square of its derivative over the interval $(0, 1)$ under the restrictions*

$$u(0) = u(1) = 0, \qquad \int_0^1 u(x)\, dx = 1.$$

We can alternatively write, as pointed out earlier,

$$u(0) = 0, \quad \int_0^1 u'(x)\,dx = 0, \quad \int_0^1 u(x)\,dx = 1.$$

The augmented integrand is now

$$\tilde{F}(x, \lambda, \xi) = \xi^2 + z_1\xi + z_2\lambda,$$

with E-L equation

$$u''(x) = \frac{z_2}{2}.$$

A first integration yields

$$u'(x) = \frac{z_2}{2}x + c,$$

and a further integration, bearing in mind that $u(0) = 0$, leads to

$$u(x) = \frac{z_2}{4}x^2 + cx.$$

Since the function $\tilde{F}$ is always strictly convex (with respect to ξ), the unique optimal solution is found by imposing the two integral constraints on the above function u, namely,

$$0 = \int_0^1 \left(\frac{z_2}{2}x + c\right) dx, \quad 1 = \int_0^1 \left(\frac{z_2}{4}x^2 + cx\right) dx.$$

After going through the computations, we obtain the optimal solution

$$u(x) = -6x(1 - x).$$

Example 5.15   (The hanging cable) A more interesting example of a variational problem under integral constraints is the following. We would like to determine the profile adopted by a uniform cable hanging from its two endpoints at the same height under the action of its own weight, assuming that this profile is the result of a minimization process of the potential energy (see Figure 5.6).

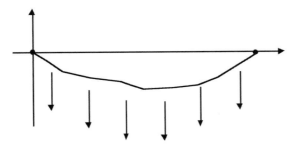

Figure 5.6. The hanging cable.

Suppose that we place the x-axis along the two points at the same height a distance D apart from each other. The length of the cable is L. Obviously, $L \geq D$. If w represents the weight per unit length, assuming that cross sections are uniform along the cable, the potential energy associated with the total weight is given by the integral

$$I(u) = w \int_0^D u(x)\sqrt{1 + u'(x)^2}\, dx,$$

where as usual,

$$ds = \sqrt{1 + u'(x)^2}\, dx$$

represents the infinitesimal element of arc length. Notice that in this case, u is to be taken negative, since we will minimize $I(u)$ among all negative functions such that $u(0) = u(D) = 0$.

There is also an important restriction to be taken into account. This is the constraint saying that the length of the cable must be L. Otherwise, the variational problem would not have any physical significance, since we could make $I(u)$ as small as possible by making u more and more negative. The constraint we are referring to is

$$L = \int_0^D \sqrt{1 + u'(x)^2}\, dx.$$

Altogether, we are willing to find optimal solutions for the variational problem

$$\text{Minimize} \quad \int_0^D u(x)\sqrt{1 + u'(x)^2}\, dx$$

subject to

$$u(0) = u(D) = 0, \quad L = \int_0^D \sqrt{1 + u'(x)^2} \, dx.$$

According to our previous discussion, we are now concerned with the integrand

$$\tilde{F}(x, \lambda, \xi) = \lambda\sqrt{1 + \xi^2} + z\sqrt{1 + \xi^2} = (\lambda + z)\sqrt{1 + \xi^2},$$

where z is the multiplier associated with the integral constraint and is regarded as a parameter. Since we formally obtain the same type of integrand as in the case of minimal surfaces of revolution, it is not difficult to check that computations are formally the same, and we arrive at the optimal solution

$$u(x) = c\cosh\left(\frac{x - D/2}{c}\right) - z,$$

where the constant c and the multiplier z are determined by the conditions

$$z = c\cosh(D/(2c)), \quad L = \int_0^D \sqrt{1 + \sinh^2\left(\frac{x - D/2}{c}\right)} \, dx.$$

The solution is therefore a catenary curve.

Example 5.16   (The channel) According to our discussion on the design of a channel in Chapter 1, the question is to determine the profile of the cross section (a curve) that encloses a fixed area has a minimum perimeter.

If u, defined in $(0, 1)$, describes one such feasible profile, we must demand

$$u(0) = u(1) = 0, \quad A = \int_0^1 u(x) \, dx,$$

and among all such curves we are seeking the one providing the least value for

$$I(u) = \int_0^1 \sqrt{1 + u'(x)^2} \, dx.$$

We assume that $u \geq 0$.

As before, we must work with the E-L equation for the function

$$\tilde{F}(x, \lambda, \xi) = \sqrt{1 + \xi^2} + z\lambda,$$

where z is the multiplier. We notice that this function is convex in (λ, ξ), affine in λ, and stictly convex in ξ. This suffices to guarantee uniqueness of the optimal solution (review the proof of Theorem 5.1). The E-L equation has the form

$$\left(\frac{u'(x)}{\sqrt{1 + u'(x)^2}} \right)' = z.$$

After a few elementary computations we have

$$u'(x) = \frac{zx + c}{\sqrt{1 - (zx + c)^2}},$$

where c is a constant. Then

$$\begin{aligned} u(x) &= \int_0^x \frac{zs + c}{\sqrt{1 - (zs + c)^2}} \, ds \\ &= -\frac{1}{z} \sqrt{1 - (zs + c)^2} \Big|_0^x \\ &= \frac{1}{z} \left(\sqrt{1 - c^2} - \sqrt{1 - (zx + c)^2} \right). \end{aligned}$$

The condition $u(1) = 0$ leads to $c = -z/2$ and hence

$$u(x) = \frac{1}{2z} \left(\sqrt{4 - z^2} - \sqrt{4 - z^2(2x - 1)^2} \right).$$

The multiplier z is to be determined to satisfy the integral constraint, and this leads to some cumbersome computations. What is important to realize is that this optimal solution u is an arc of a circle.

We now treat the case of integral constraints in the form of equalities and inequalities simultaneously. We focus on the problem

$$\text{Minimize} \quad I(u) = \int_a^b F(x, u(x), u'(x)) \, dx$$

subject to

$$u(a) = A,$$

$$\int_a^b G(x, u(x), u'(x)) \, dx \le \alpha,$$

$$\int_a^b H(x, u(x), u'(x)) \, dx = \beta.$$

Again, our experience with nonlinear programming makes the following result plausible.

Proposition 5.17   *Suppose there is a vector (y, z), $y \ge 0$, such that the function*

$$\tilde{F}(x, \lambda, \xi) = F(x, \lambda, \xi) + yG(x, \lambda, \xi) + zH(x, \lambda, \xi)$$

is convex in (λ, ξ) (again the convexity with respect to ξ suffices.) If v is the (unique) solution of the corresponding E-L problem,

$$\frac{d}{dx}\left[\tilde{F}_\xi(x, v(x), v'(x)) \right] = \tilde{F}_\lambda(x, v(x), v'(x)), \quad v(a) = A,$$

then v is the optimal solution of the above variational problem, provided that

$$\int_a^b G(x, v(x), v'(x)) \, dx \le \alpha,$$

$$\int_a^b H(x, v(x), v'(x)) \, dx = \beta,$$

$$y\left(\int_a^b G(x, v(x), v'(x)) \, dx - \alpha \right) = 0.$$

Let us examine one clarifying example.

Example 5.18   *Solve*

$$\text{Minimize} \quad \int_0^1 u'(x)^2 \, dx$$

subject to

$$u(0) = 0, u(1) = 1, \quad \int_0^1 u(x)^2 \, dx \le \alpha,$$

where α is a given nonnegative number. Introducing a multiplier y to take care of the integral constraint, we must face the E-L equation for the augmented integrand

$$\tilde{F}(x, \lambda, \xi) = \xi^2 + y\lambda^2,$$

regarding y as a parameter. This problem consists of

$$u''(x) = yu(x), \quad u(0) = 0, \quad u(1) = 1.$$

The general solution is of the form

$$u(x) = \frac{\sinh(\sqrt{y}\,x)}{\sinh(\sqrt{y})}$$

if y does not vanish, and

$$u(x) = x$$

if $y = 0$. Whenever α is chosen so that

$$\int_0^1 x^2\,dx = \frac{1}{3} \le \alpha,$$

the optimal solution will be the line $u(x) = x$. But if

$$\alpha < \frac{1}{3},$$

the optimal solution will be of the form

$$u(x) = \frac{\sinh(\sqrt{y}\,x)}{\sinh(\sqrt{y})},$$

where the multiplier y is determined so that

$$\int_0^1 \left(\frac{\sinh(\sqrt{y}\,x)}{\sinh(\sqrt{y})} \right)^2 dx = \alpha.$$

When additional restrictions for variational problems come in a pointwise fashion like

$$H(x, u(x), u'(x)) = 0$$

or

$$G(x, u(x), u'(x)) \leq 0, \quad H(x, u(x), u'(x)) = 0,$$

then multipliers $y(x), z(x)$ must be functions of x, because we have to satisfy a constraint for each x. In this way the functional that lets us treat this sort of pointwise constraint is

$$I(u, y, z) = \int_a^b [F(x, u(x), u'(x)) + y(x)G(x, u(x), u'(x))$$
$$+ z(x)H(x, u(x), u'(x))] \, dx.$$

Notice the explicit dependence of I with respect to the multipliers $y(x), z(x)$.

There is no doubt that one of the most important situations in which pointwise constraints must be taken into account is that of optimal control problems. Since the last chapter of this text is devoted, because of its importance, to this class of problems, we will not say anything else at this stage.

6. SUMMARY OF RESTRICTIONS FOR VARIATIONAL PROBLEMS

This section is intended to clarify all the possibilities that may arise in considering a typical variational problem from the perspective of different types of constraints. This discussion incorporates, as particular cases, transversality conditions of all kinds, and all situations related to integral constraints and restrictions on endpoints.

Consider the problem

$$\text{Minimize} \quad \int_a^b F(x, u(x), u'(x)) \, dx$$

subject to

$$\int_a^b G(x, u(x), u'(x)) \, dx \leq \alpha, \quad \int_a^b H(x, u(x), u'(x)) \, dx = \beta, \quad u(c) \leq A,$$

where $c = a$ or $c = b$ is fixed.

We know that an optimal solution must solve E-L for the modified integrand

$$\tilde{F} = F + yG + zH.$$

Assume that $\tilde{u}(x, y, z, w)$ is the general solution of the E-L equation for $\tilde{F}$, where $w = (w_1, w_2)$ represents two arbitrary constants of integration, since the E-L equation is of second order. Define

$$f(y, z, w) = \int_a^b F(x, \tilde{u}(x, y, z, w), \tilde{u}'(x, y, z, w))\, dx,$$

$$g(y, z, w) = \int_a^b G(x, \tilde{u}(x, y, z, w), \tilde{u}'(x, y, z, w))\, dx,$$

$$h(y, z, w) = \int_a^b H(x, \tilde{u}(x, y, z, w), \tilde{u}'(x, y, z, w))\, dx,$$

$$\varphi(y, z, w) = \tilde{u}(c, y, z, w).$$

Then it is not hard to convince ourselves that the optimal values for (y, z, w) ought to be detemined by solving the NLPP

$$\text{Minimize} \quad f(y, z, w)$$

subject to

$$g(y, z, w) \leq \alpha, \quad y \geq 0, \quad y\,(g(y, z, w) - \alpha) = 0,$$
$$h(y, z, w) = \beta, \quad \varphi(y, z, w) \leq A.$$

Once these optimal values (y_0, z_0, w_0) have been found, it is important to go back to $\tilde{F}$ "a posteriori" and check that

$$\tilde{F} = F + y_0 G + z_0 H$$

is convex with respect to ξ. In particular, since $y \geq 0$, the function G (or all of its components) must be convex in ξ. If $\tilde{F}$ is not convex, then we may not have the optimal solution. In this framework we may also treat all variants related to different types of inequalities and/or equalities.

This general perspective sometimes leads to nonsmooth or even in some cases to noncontinuous problems that would require, in principle, more elaborate techniques to find optimal solutions. Because of the special structure of constraints, it is elementary to check that, as we pointed out when dealing with optimality conditions for NLPP, the restrictions

$$g(y, z, w) \leq \alpha, \quad y \geq 0, \quad y\,(g(y, z, w) - \alpha) = 0$$

amount to having

$$y_i \left(g_i(y, z, w) - \alpha_i \right) = 0$$

for all i, and these equations allow a separate treatment of several NLPP. For instance, when G has a single component, then we would have two possibilities,

$$y = 0 \quad \text{or} \quad g(y, z, w) = \alpha,$$

leading to the two NLPP

$$\text{Minimize} \quad f(0, z, w)$$

subject to

$$g(0, z, w) \le \alpha, \quad h(0, z, w) = \beta, \quad \varphi(0, z, w) \le A;$$

and

$$\text{Minimize} \quad f(y, z, w)$$

subject to

$$g(y, z, w) = \alpha, \quad h(y, z, w) = \beta, \quad \varphi(y, z, w) \le A,$$

considering those optimal solutions for $y \ge 0$. The true optimal solution of our problem will be found in one of these two NLPP.

Example 5.19   *We would like to*

$$\text{Minimize} \quad \frac{1}{2} \int_0^1 \left(u'(x) - 1 \right)^2 dx$$

subject to

$$0 \le u(0) \le 1, \quad 0 \le u(1) \le \frac{1}{2}.$$

The E-L equation is $u'' = 0$, so that the general solution is

$$u(x) = w_1 x + w_2.$$

Therefore, we consider the objective function

$$f(w) = \frac{1}{2} \int_0^1 (w_1 - 1)^2 dx = \frac{1}{2}(w_1 - 1)^2.$$

Restrictions correspond to requiring

$$0 \le u(0) = w_2 \le 1, \quad 0 \le u(1) = w_1 + w_2 \le \frac{1}{2}.$$

Hence we must solve the NLPP

$$\text{Minimize} \quad \frac{1}{2}(w_1 - 1)^2$$

under the constraints

$$0 \le w_1 \le 1, \quad 0 \le w_1 + w_2 \le \frac{1}{2}.$$

It is very easy to find (even graphically) that the optimal solution corresponds to $(1/2, 0)$, so that the optimal solution of our initial problem is $u(x) = x/2$.

Example 5.20   Consider the problem

$$\text{Minimize} \quad \frac{1}{2}\int_0^1 u'(x)^2 \, dx$$

under the constraints

$$0 \le u(0) \le 1, \quad u(1) = 1, \quad \int_0^1 u(x) \, dx \le \frac{1}{2}.$$

The E-L equation we should solve is $u'' = y$, whose general solution is

$$u(x) = y\frac{x^2}{2} + w_1 x + w_2.$$

It is very easy to obtain

$$f(y, w) = \frac{1}{6}\left(y^2 + 3yw_1 + 3w_1^2\right),$$
$$g(y, w) = \frac{y}{6} + \frac{w_1}{2} + w_2,$$
$$\varphi(y, w) = \left(w_2, w_1 + w_2 + \frac{y}{2}\right).$$

Thus the NLPP to be considered is

$$\text{Minimize} \quad \frac{1}{6}\left(y^2 + 3yw_1 + 3w_1^2\right)$$

subject to

$$\frac{y}{6} + \frac{w_1}{2} + w_2 \leq \frac{1}{2}, \quad w_1 + w_2 + \frac{y}{2} = 1, \quad 0 \leq w_2 \leq 1,$$

$$y \geq 0, \quad y\left(\frac{y}{6} + \frac{w_1}{2} + w_2 - \frac{1}{2}\right) = 0.$$

As remarked earlier, this NLPP splits into

$$\text{Minimize} \quad \frac{w_1^2}{2}$$

subject to

$$\frac{w_1}{2} + w_2 \leq \frac{1}{2}, \quad w_1 + w_2 = 1, \quad 0 \leq w_2 \leq 1,$$

with $(0,1)$ as unique admissible point and associated cost $1/2$; and

$$\text{Minimize} \quad \frac{1}{6}\left(y^2 + 3yw_1 + 3w_1^2\right)$$

subject to

$$\frac{y}{6} + \frac{w_1}{2} + w_2 = \frac{1}{2}, \quad w_1 + w_2 + \frac{y}{2} = 1, \quad 0 \leq w_2 \leq 1,$$

with $y \geq 0$. Using the two linear constraints we obtain $w_1 = 1 - 2y/3$, $w_2 = y/6$, and the condition $0 \leq w_2 \leq 1$ leads to $0 \leq y \leq 6$. Altogether, and making the appropriate substitutions, we are interested in the minimum of the parabola

$$\frac{1}{6}\left(\frac{y^2}{3} - y + 3\right)$$

over the interval $0 \leq y \leq 6$. Such a minimum is attained at $y = 3/2$ with cost $3/8$ and $w_1 = 0$, $w_2 = 1/4$. Since this optimal cost is smaller than the one found in the previous subproblem, we conclude that the optimal solution sought is

$$u(x) = \frac{3x^2 + 1}{4}.$$

7. VARIATIONAL PROBLEMS OF DIFFERENT ORDER

In some instances we may be interested in variational problems where either second (or higher) derivatives appear or else no derivative at all is present. The highest order of derivatives appearing in a variational problems is the order of the problem. So far, we have focused on first order problems. These are by far the most common. But we would like to say a few words about variational problems of zero and second order.

Variational problems of order zero, i.e., no derivatives appear in the cost functional, are in fact included in our previous discussion, because after all, these are a special case of first order problems where there is no dependence on first derivatives. In such cases, the E-L equation is no longer a differential equation but rather an algebraic equation. Solving this equation will provide the optimal solution after the adjustment of constants with endpoint conditions. An example will clarify what we mean.

Example 5.21   *A cylindrical container rotates around its axis at constant angular velocity ω_0 (Figure 5.7). We would like to determine the profile adopted by a certain fluid in its interior, assuming that the shape is the result of a minimization process of potential energy.*

Specifically, and due to radial symmetry, if

$$z(r), \quad 0 \leq r \leq R,$$

describes a given profile, the potential energy associated with it is expressed through the integral

$$U(z) = \int_0^R \pi \rho \left[gr \left(z(r)^2 - 2Hz(r) \right) - \omega_0^2 r^3 z(r) \right] dr,$$

where g, ρ, and H are constants. Obviously, the profile $z(r)$ must respect the volume constraint

$$V = 2\pi \int_0^R rz(r) \, dr$$

for a fixed constant V. In summary, we seek to

$$\text{Minimize} \quad \int_0^R \left[gr \left(z(r)^2 - 2Hz(r) \right) - \omega_0^2 r^3 z(r) \right] dr$$

subject to

$$V = 2\pi \int_0^R rz(r)\,dr.$$

Notice that there is no derivative $z'(r)$ appearing in the cost functional.

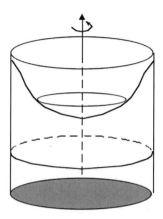

Figure 5.7. The profile of a rotating fluid.

If we introduce a multiplier λ to take into account the volume constraint, we have to write down the E-L equation for the function

$$F(x, u) = \left[g(u^2 - 2Fu)r - \omega_0^2 r^3 u\right] + \lambda r u.$$

This is now

$$0 = \frac{\partial F}{\partial u}(r, z(r)),$$

i.e.,

$$g\left(2z(r) - 2H\right)r - \omega_0^2 r^3 + \lambda r = 0.$$

Solving for $z(r)$, we obtain

$$z(r) = \frac{\omega_0^2}{2g}r^2 + H - \frac{\lambda}{2g}.$$

By means of the volume constraint we determine the constant

$$H - \frac{\lambda}{2g}.$$

Indeed, we have

$$H - \frac{\lambda}{2g} = \frac{V}{\pi R^2} - \frac{\omega_0^2}{4g} R^2,$$

and the optimal profile is

$$z(r) = \frac{\omega_0^2}{2g} \left(r^2 - \frac{R^2}{2} \right) + \frac{V}{\pi R^2}.$$

Notice that this is a parabola rotated around the axis of the cylinder.

Second order problems are more involved, as one would reasonably anticipate. The strategy for finding optimality conditions in the form of E-L equations
is, however, similar to the case of first order problems. The E-L equation will
be a fourth order differential equation, which will be complemented with appropriate endpoint or boundary conditions. We leave it to the interested reader
(although we actually include such discussion below) to justify the following
fact.

Theorem 5.22 *If the integrand for a second order problem is*

$$F(x, u, u', u''),$$

then the E-L equation reads

$$\frac{d^2}{dx^2} \frac{\partial F}{\partial u''} - \frac{d}{dx} \frac{\partial F}{\partial u'} + \frac{\partial F}{\partial u} = 0.$$

Endpoint conditions may involve values of u and/or u'. Transversality or
natural boundary conditions will have to be kept in mind as well.

Many problems related to bending and buckling of thin elastic rods are
formulated by means of second order variational problems, because energies
depend on curvatures (second derivatives) of adopted profiles. One elementary
example follows.

Example 5.23 *A thin elastic rod is deflected as shown in Figure 5.8. If*

$$u(x), \quad x \in (0, L),$$

describes the center line of the rod, the potential energy accumulated in such a state is given by the integral

$$P(u) = k \int_0^L \frac{u''(x)^2}{(1 + u'(x)^2)^{5/2}} \, dx,$$

where $k > 0$ is a known constant. Endpoint boundary conditions are

$$u(0) = u'(0) = 0, \quad u(L) = L_1.$$

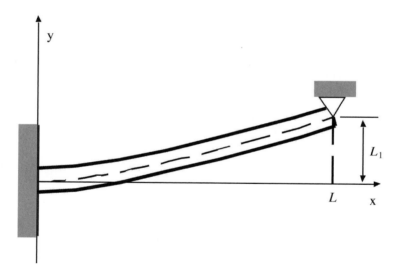

Figure 5.8. Deflection of a thin elastic rod.

Since we have three imposed conditions on endpoints, we need another one, since the E-L equation will have order four. This missing condition is a natural boundary condition at the right endpoint L. Since the missing boundary condition is $u'(L)$, the transversality condition we need is the factor going with $u'(L)$ when we integrate by parts in analyzing the function

$$g(t) = \int_0^L F(x, u(x) + t\varphi(x), u'(x) + t\varphi'(x), u''(x) + t\varphi''(x))\, dx,$$

and demanding

$$0 = g'(0) = \int_0^L \left(\frac{\partial F}{\partial u}\varphi + \frac{\partial F}{\partial u'}\varphi' + \frac{\partial F}{\partial u''}\varphi'' \right) dx.$$

By integrating twice by parts and bearing in mind the other endpoint conditions, we obtain

$$0 = \int_0^L \varphi \left(\frac{d^2}{dx^2}\frac{\partial F}{\partial u''} - \frac{d}{dx}\frac{\partial F}{\partial u'} + \frac{\partial F}{\partial u} \right) dx + \varphi'(L)\frac{\partial F}{\partial u''}(L, u(L), u'(L), u''(L)).$$

This informs us about the E-L equation, and also tells us that the natural boundary condition we are seeking is

$$\frac{\partial F}{\partial u''}(L, L_1, u'(L), u''(L)) = 0.$$

In our particular example, this condition is

$$u''(L) = 0.$$

After writing carefully E-L, we must solve the problem

$$\frac{d^2}{dx^2}\frac{2u''}{(1 + u'^2)^{5/2}} + \frac{d}{dx}\frac{5(u'')^2 u'}{(1 + u'^2)^{7/2}} = 0,$$
$$u(0) = u'(0) = 0, \quad u(L) = L_1, \quad u''(L) = 0.$$

This equation is impossible to solve by hand. A reasonable approximation would indicate that the expected profile will have a small derivative u', so that we can neglect terms involving u'. This simplification, together with one integration, leads to

$$u''' + \frac{5}{2}(u'')^2 u' = \text{constant},$$

and even further,

$$u''' = \text{constant}.$$

This together with the four boundary conditions implies that

$$u(x) = -\frac{L_1}{2L^3}x^3 + \frac{3L_1}{2L^2}x^2$$

provides a reasonable approximation of the profile adopted by the rod.

8. DYNAMIC PROGRAMMING: BELLMAN'S EQUATION

In many practical situations of interest, a system is to move succesively through a number of different steps to complete a whole process and to arrive to a desired state. Each one of those actions has an associated cost. Given a specific objective to be reached from a given initial state, we would like to determine the optimal global strategy that has an associated least cost.

Let t denote the variable indicating the succesive stages in which a decision must be made about where to lead the system

$$t = t_i, \quad i = 0, 1, \ldots, n;$$

let x be the variable describing the state of the system. At each step i, we ought to have

$$x \in A_i$$

if A_i is the (finite) set of feasible states when $t = t_i$. The cost associated with the passage from $x \in A_i$ to $y \in A_{i+1}$ is denoted by

$$c(i, x, y).$$

Given an initial state (t_0, x_0), we are concerned with the task of determining the optimal strategy to reach the final desired state (t_n, x_n) with the least cost. This is a typical situation of dynamic programming.

Assume that for $0 < j < n$ we know the optimal path starting at (t_0, x_0) and going to (t_j, x) for each $x \in A_j$. Let

$$S(t_j, x)$$

provide the cost associated with such an optimal strategy ending at (t_j, x). How could we find the optimal solution starting at (t_0, x_0) and ending at (t_{j+1}, y) for any given $y \in A_{j+1}$? It is not hard to be convinced that we should solve the problem

$$\min_{x \in A_j} [S(t_j, x) + c(j, x, y)].$$

This is the fundamental law or property of dynamic programming, and through it we can find the optimal strategy from (t_0, x_0) to (t_n, x_n) in the most rational way.

Proposition 5.24 *(Fundamental property of dynamic programming) If $S(t_j, x)$ denotes the optimal cost from (t_0, x_0) to (t_j, x), then we must have*

$$S(t_{j+1}, y) = \min_{x \in A_j} [S(t_j, x) + c(j, x, y)].$$

A typical simplified situation follows.

Example 5.25 *(The traveler) One passenger wants to go from city A to city H through the shortest path according to the map in Figure 5.9, where numbers indicate distances between corresponding cities.*

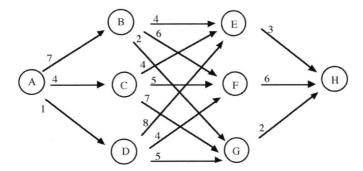

Figure 5.9. The traveler.

Evidently, we could make an exhaustive counting of all possibilities and decide on the best one. In this simplified situation this would not be a bad idea. However, we would like to illustrate the fundamental property of dynamic *programming in this example. According to Proposition 5.24, we must proceed succesively to determine $S(t_j, x)$ for each $x \in A_j$ to end with $S(t_n, x_n)$. In the proposed example, we have four stages t_0, t_1, t_2, t_3 with associated sets of feasible states*

$$A_0 = \{A\}, \quad A_1 = \{B, C, D\}, \quad A_2 = \{E, F, G\}, \quad A_3 = \{H\}.$$

For each city in A_1 there is a unique path from A, so that it must be optimal, and
$$S(t_1, B) = 7, \quad S(t_1, C) = 4, \quad S(t_1, D) = 1.$$

For each city in A_2, we determine the optimal cost based on the fundamental property of dynamic programming,

$$S(t_{j+1}, y) = \min_{x \in A_j} \left[S(t_j, x) + c(j, x, y) \right].$$

In our concrete example we are looking for the minimum of

$$
\begin{array}{ccc}
7 + 4, & 4 + 4, & 1 + 8, \\
7 + 6, & 4 + 5, & 1 + 4, \\
7 + 2, & 4 + 7, & 1 + 5,
\end{array}
$$

to find that
$$S(t_2, E) = 8, \quad S(t_2, F) = 5, \quad S(t_2, G) = 6.$$

The last step leads us to the desired shortest path. We ought to decide the minimum of
$$\min \{8 + 3, 5 + 6, 6 + 2\} = 8.$$

Consequently, the shortest distance is 8, and it corresponds to the route $A - D - G - H$. Note that in using the fundamental property of dynamic programming we always have to operate with sums of two numbers, while a direct exhaustive counting would require (in this simple situation) to work with sums of three numbers. It is not difficult to infer the importance of this fact for more complicated situations.

Example 5.26   *Another typical situation in discrete dynamic programming concerns a company that can produce three different food products from milk:*

cheese, butter, and yogurt. The benefit from these products using 1, 2, 3, or 4 units of milk is given in Table 5.10.

What is the maximum benefit that can be obtained with 4 units of milk? Again, in this simplified situation, it would not be hard to find the solution by exhaustively examining all possibilities. The scheme in the context of dynamic programming would be as follows: We identify the three dairy products with the values of the variable t, t_0, t_1, and t_2. The set of possible states in each step will be $\{0, 1, 2, 3, 4\}$, meaning that we can assign each one of those numbers of units of milk to each of the three products, as long as we do not exceed the available 4 units. By using the data in Table 5.10, we get the results of Table 5.11.

Number of units of milk	1	2	3	4
Cheese	8	18	22	24
Butter	3	6	9	12
Yogurt	6	7	8	10

Table 5.10. Individual benefit from dairy products.

The maximum benefit is 28, and it corresponds to 3 units of milk for cheese and 1 unit for yogurt.

The basic principle of dynamic programming applied to the continuous case furnishes another perspective on variational problems in which we focus on the optimal values rather than on optimal solutions. We define the "value function" by putting

$$S(t, x) = \min_u \left\{ \int_t^T F(\tau, u(\tau), u'(\tau))\, d\tau : u(t) = x, u(T) = B \right\},$$

where T and B are fixed given data. Then $S(t, x)$ yields the optimal cost associated with the problem starting at (t, x) and ending at (T, B). In the variational approach, we insisted on optimal paths or solutions, and not so much on optimal values.

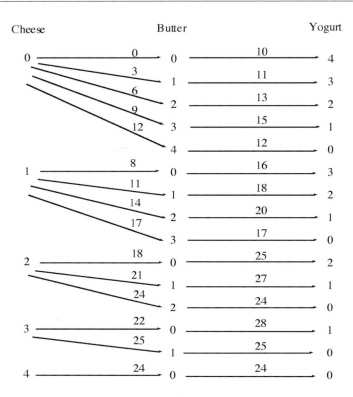

Table 5.11. Benefits from dairy products.

The basic property of this value function $S(t, x)$ is precisely the fundamental principle of dynamic programming, already discussed in the discrete case, which in this context can be written as

$$S(t, x) = \min_{z} \left\{ \min_{u} \left\{ \int_{t}^{t'} F(\tau, u(\tau), u'(\tau)) \, d\tau : u(t) = x, u(t') = z \right\} + S(t', z) \right\},$$

where $t < t' < T$ (see Figure 5.12).

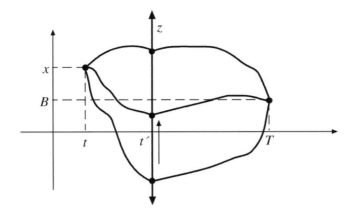

Figure 5.12. The fundamental property of the value function.

This condition can be reorganized as follows:

$$0 = \min_z \left\{ \min_u \left\{ \frac{1}{t' - t} \int_t^{t'} F(\tau, u(\tau), u'(\tau)) \, d\tau : u(t) = x, u(t') = z \right\} \right. $$
$$\left. + \frac{S(t', z) - S(t, x)}{t' - t} \right\}.$$

If we let $z = x + y(t' - t)$, the minimum on z can be transformed into a minimum on y, and

$$0 = \min_y \left\{ \min_u \left\{ \frac{1}{t - t'} \int_t^{t'} F(\tau, u(\tau), u'(\tau)) \, d\tau : u(t) = x, u(t') = x + y(t' - t) \right\} \right.$$
$$\left. + \frac{S(t', x + y(t' - t)) - S(t, x)}{t - t'} \right\}.$$

What happens if we let $t' \searrow t$? For each function u such that $u(t) = x$ and $u(t') = x + y(t' - t)$, by the fundamental theorem of calculus, we have

$$\frac{1}{t' - t} \int_t^{t'} F(\tau, u(\tau), u'(\tau)) \, d\tau \to F(t, x, y),$$

while if we assume that S is differentiable, then by the chain rule,

$$\frac{S(t', x + y(t' - t)) - S(t, x)}{t' - t} \rightarrow \frac{\partial S}{\partial t}(t, x) + y\frac{\partial S}{\partial x}(t, x).$$

We conclude that

$$0 = \min_y \left[F(t, x, y) + \frac{\partial S}{\partial t}(t, x) + y\frac{\partial S}{\partial x}(t, x) \right],$$

or

$$-\frac{\partial S}{\partial t}(t, x) = \min_y \left[F(t, x, y) + y\frac{\partial S}{\partial x}(t, x) \right].$$

This is Bellman's equation of dynamic programming. We have included this informal derivation of it because we will follow a similar path to establish Pontryagin's maximum principle for optimal control problems in the next chapter.

This approach may lead, without great difficulty, to the E-L equation for optimal solutions of variational problems. Since as just remarked, this will be our main strategy for necessary conditions of optimality in the next chapter, we do not insist on it here.

Example 5.27   *In some simple situations, the computations to make Bellman's equation explicit can be carried out. Assume that $F(t, \lambda, \xi) = \xi^2$. Since this function is strictly convex and it depends only on the derivative, we know that the optimal solution u satisfying $u(t) = x$, $u(T) = B$ is the linear function with slope $(B - x)/(T - t)$. Therefore, the value function is*

$$S(t, x) = (T - t)\left(\frac{B - x}{T - t}\right)^2 = \frac{(B - x)^2}{T - t}.$$

On the other hand, for any constant α,

$$\min_y (y^2 + y\alpha) = -\frac{\alpha^2}{4},$$

so that Bellman's equation is

$$4\frac{\partial S}{\partial t} = \left(\frac{\partial S}{\partial x}\right)^2.$$

It is a simple exercise to check that the explicit form for $S(t, x)$ does indeed satisfy this partial differential equation.

9. SOME BASIC IDEAS ON THE NUMERICAL APPROXIMATION

It is almost evident that optimal solutions for many problems cannot be found analytically. Approximation techniques are thus an indispensable tool for solving many variational problems. The basic idea of numerical approximation of continuous optimization problems is "'discretization."

Given a variational problem, we must build a discretized version of it with a certain level of accuracy that is related to the fineness of the discretization we have utilized. Such a discretized version will now be a programming problem, so that we can apply all the computational algorithms described in Chapter 4. From this point of view, numerical algorithms are the common link between finite and infinite dimensional optimization problems.

Another important possibility for approximating optimal solutions of variational problems is to exploit E-L equations. But since this is a book about optimization, and that other approach will lead us to approximate differential equations, we will stick to genuine optimization techniques and concepts. It is, however, important always to bear in mind any valuable information about the problem whose solution we seek to approximate.

Suppose we have a typical variational problem

$$\text{Minimize} \quad I(u) = \int_a^b F(x, u(x), u'(x)) \, dx$$

with $u(a) = A$, $u(b) = B$. Usually, discretizations of integrals are set up by dividing the interval of integration $[a, b]$ into a certain number of subintervals, $n + 1$, and assume that feasible functions for the new, discretized, optimization problem are piecewise affine, i.e., they are affine on each subinterval

$$\left[a + j \frac{(b - a)}{(n + 1)}, a + (j + 1) \frac{(b - a)}{(n + 1)} \right]. \tag{5-2}$$

Notice that such functions are uniquely determined by their values at the nodes

$$a + j \frac{(b - a)}{(n + 1)}, \quad j = 1, \ldots, n,$$

and therefore feasible vectors for the new optimization problem will correspond to these values. We see that this process will change the original infinite dimensional problem to a finite dimensional one. The point is that by letting $n + 1$, the number of subintervals, become larger and larger, optimal solutions for these discretized optimization problems will resemble and approximate fairly well, under conditions we will overlook here, the true optimal solutions for the initial optimization problem. Let

$$X = (x_j)_{1 \leq j \leq n} \tag{5–3}$$

be the nodal values of feasible functions. In this way the function u that we will consider for $I(u)$ will be

$$u(x) = A + \sum_{k=1}^{j} \frac{(x_{k+1} - x_k)}{n+1} + (x_{j+1} - x_j) \left(x - a - k \frac{b-a}{n+1} \right)$$

if

$$x \in \left[a + k \frac{b-a}{n+1}, a + (k+1) \frac{b-a}{n+1} \right].$$

This is the continuous, piecewise-affine function that takes on values x_j at the nodal points $a + j(b - a)/(n + 1)$. There is a useful way of expressing this function as a linear combination of certain "basic functions". Namely, if we define

$$\psi_{j,n}(x), \quad j = 0, 1, \ldots, n, n + 1,$$

as the piecewise affine function whose value at nodes x_i for $i \neq j$ is zero and precisely at x_j is unity (see Figure 5.13), then it is clear that the piecewise-linear function u with nodal values x_j can be written as

$$u(x) = \sum_{j} x_j \psi_{j,n}(x). \tag{5–4}$$

For the derivative, we have

$$u'(x) = \sum_{j} x_j \psi'_{j,n}(x).$$

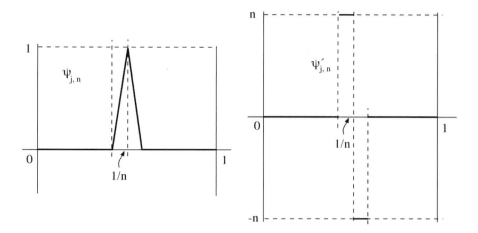

Figure 5.13. Basic functions for the numerical approximation.

In this way, it is immediate how to write an optimization problem for the vector X in (5–3) by computing $I(u)$ for u as in (5–4). This will be the discrete, approximated version of our continuous variational problem. Specifically,

$$T(X) = I(u) = \int_a^b F(x, u(x), u'(x))\, dx$$

$$= \int_a^b F\left(x, \sum_j x_j \psi_{j,n}(x), \sum_j x_j \psi'_{j,n}(x)\right) dx.$$

If we break this integral into a sum over the subintervals (5–2), and realize that each $\psi_{j,n}$ is linear, so that its derivative is constant on them, gathering all contributions on a particular subinterval, we can more explicitly write

$$T(X) = \sum_{j=0}^n \int_{a+j(b-a)/(n+1)}^{a+(j+1)(b-a)/(n+1)} F\left(x, \sum_j x_j \psi_{j,n}(x), (n+1)(x_{j+1} - x_j)\right) dx.$$

Even further, we can use a simple quadrature rule to approximate these integrals. The final form, by using the trapezoidal rule, is

$$T(X) = \sum_{j=0}^{n} \frac{1}{2(n+1)} \left[F\left(a + (j+1)\frac{b-a}{n+1}, x_{j+1}, (n+1)(x_{j+1} - x_j) \right) \right.$$
$$\left. + F\left(a + j\frac{b-a}{n+1}, x_j, (n+1)(x_{j+1} - x_j) \right) \right],$$

where $x_0 = A$, $x_{n+1} = B$. In terms of the vector X of nodal values, we are faced with a (nonlinear, unconstrained) programming problem. By solving it, we obtain an approximate solution to the initial, continuous optimization problem. The form of the functional $T(X) = I(u)$ in terms of X depends on the particular situation.

Example 5.28 *For the minimal surface of revolution example we take*

$$I(u) = \int_0^1 u(x)\sqrt{1 + u'(x)^2}\, dx,$$
$$u(0) = 1, \quad u(1) = 1.$$

The resulting objective function $T(X)$ in terms of the nodal values, as pointed out in the above discussion, is

$$T(X) = \sum_{j=0}^{n} \sqrt{1 + ((n+1)(x_{j+1} - x_j))^2}\, \frac{x_{j+1} + x_j}{2(n+1)}.$$

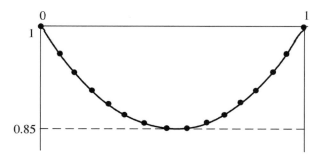

Figure 5.14. Approximation of the minimal surface of revolution.

Having in mind $x_0 = 1$, $x_{n+1} = 1$, we see that it is this functional that we would like to minimize with the help of some of the algorithms in Chapter 4, for several values of the number of subintervals n. By doing so, we find very good agreement with the arc of a catenary, which is the optimal solution of the continuous optimization problem (Figure 5.14).

Alternatively, we can set up the discretization scheme considering as independent variables the slopes on each subinterval. This will lead to a simpler form of the objective functional, but we would have to enforce a (linear) constraint, because the slopes in the different subintervals must be such that the value of u at 1 is given, and this imposes a restriction on the set of possible slopes. Rather than solving the same example in this format with an integral constraint, we prefer to start the next example.

Example 5.29 *The problem for the design of a channel is*

$$\text{Minimize} \quad \int_0^1 \sqrt{1 + u'(x)^2} \, dx$$

subject to

$$u(0) = u(1) = 0, \quad \frac{1}{3} = \int_0^1 u(x) \, dx.$$

As discussed in the preceding example, by dividing the interval $[0,1]$ into $n+1$ equal subintervals with equally spaced nodes

$$\frac{j}{n+1}, \quad j = 0, 1, \ldots, n+1,$$

we end up with a cost functional

$$T(X) = \frac{1}{n+1} \sum_{j=0}^{n} \sqrt{1 + (n+1)^2 \left(x_{j+1} - x_j\right)^2}.$$

The constraints are $x_0 = x_{n+1} = 0$ and

$$\frac{1}{3} = \sum_{j=0}^{n} \frac{x_{j+1} + x_j}{2(n+1)}.$$

This last constraint can be rewritten in the form

$$\frac{n+1}{3} = \sum_{j=1}^{n} x_j,$$

where we have taken into account the boundary constraints $x_0 = x_{n+1} = 0$. *Altogether, we would like to find the optimal solution of*

$$\text{Minimize} \quad \sum_{j=0}^{n} \sqrt{1 + (n+1)^2 (x_{j+1} - x_j)^2}$$

subject to

$$x_0 = x_{n+1} = 0, \quad \frac{n+1}{3} = \sum_{j=1}^{n} x_j,$$

a nonlinear programming problem under linear constraints. Figure 5.15 shows the resulting approximation for a particular value of n. *Notice that we have neglected a positive multiplicative constant for the functional.*

10.   EXERCISES

1. Determine the geodesics on a sphere. Try to make a guess about what the geodesics might be, and then argue that those indeed are the ones providing the minimum value of the appropriate functional.
2. Investigate Exercise 9 of Chapter 1, trying to figure out optimal paths.
3. Write the E-L equation for the following functionals:

$$I(u) = \int \left[(u')^2 + e^u \right] dx, \quad I(u) = \int uu' \, dx, \quad I(u) = \int x^2 (u')^2 \, dx.$$

4. Find the optimal solution of the problem

$$\min \left\{ \int_0^1 \left[x'(t)^2 + 2\, x(t)^2 \right] e^t \, dt : x(0) = 0, x(1) = e - e^{-2} \right\}$$

and the value of the minimum.

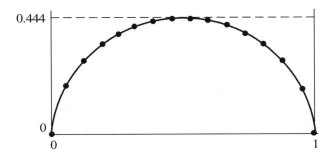

Figure 5.15. Approximation of the optimal design for the channel.

5. Solve the following variational problem:

$$\text{Minimize} \quad I(u) = \int_0^1 \left(u'(x)^2 + u(x)u'(x) + u(x)^2 \right) dx$$

among all functions satisfying $u(0) = 0$, $u(1) = 1$.

6. Solve the problem of minimizing the functional

$$I(u) = \int_0^1 u'(x)^2 \, dx, \quad u(0) = a_0, \quad u(1) = a_1,$$

among all functions satisfying

$$0 = \int_0^1 u(x)\cos(b_i x) \, dx, \quad i = 1, 2, \ldots, N,$$

where a_0, a_1, and b_i are fixed parameters.

7. Find the function $u(x)$ that minimizes

$$I(u) = \int_0^1 \left(1 + u''(x)^2 \right) dx$$

under the constraints $u(0) = 0$, $u(1) = 1$, $u'(0) = 1$, $u'(1) = 1$.

8. Solve the same question as in the preceding exercise with the functional

$$I(u) = \int_0^{\pi/2} \left(u''(x)^2 - u(x)^2 + x^2 \right) dx,$$

subject to

$$u(0) = 1, \ u(\pi/2) = 0, \ u'(0) = 0, \ u'(\pi/2) = 1.$$

9. The maximum entropy principle selects the probability distribution over the semi-infinite interval $(0, \infty)$ maximizing the integral

$$H = -\int_0^\infty u(t) \log u(t) \, dt.$$

If we impose the restrictions

$$\int_0^\infty u(t) \, dt = 1, \quad \int_0^\infty tu(t) \, dt = 1/a,$$

check that the most probable distribution is $u(t) = ae^{-at}$.

10. Consider the function

$$f(y) = \min \left\{ \int_0^{\log 2} \left[x'(t)^2 + x(t)^2 \right] dt : x(0) = 1, x(\log 2) = y \right\}.$$

Find an explicit expression for $f(y)$ and use it to determine the solution of the problem

$$\min \left\{ \int_0^{\log 2} \left[x'(t)^2 + x(t)^2 \right] dt : x(0) = 1 \right\}.$$

Solve directly this last problem and compare your results.

11. Determine the minimum value of the integrals

$$\int_0^{\log 2} \left[\left(x'(t) - 1 \right)^2 + x(t)^2 \right] dt$$

among all functions $x(t)$.

12. Solve the problem of the rope of variable cross section of Chapter 1.

    1. Show that by the change

    $$b(x) = W + \rho g \int_x^L a(s)\, ds,$$

    the problem can be reformulated by

    $$\text{Minimize} \quad \left(-\frac{\rho g}{E}\right) \int_0^L \frac{b(x)}{b'(x)}\, dx$$

    subject to

    $$b(0) = W + \rho g V, \quad b(L) = W.$$

    2. Solve the problem in this form, and interpret the final result in terms of the initial setting.

13. Sometimes, the exact formulation of a variational problem is impossible to solve, and reasonable approximations must be made. The problem of the solid moving in a bath described in Chapter 1 is one such example. Make reasonable simplifications as in Example 5.23 to obtain a good approximation to the profile of the moving solid.

14. Study the problem

    $$\text{Minimize} \quad \frac{1}{2} \int_0^1 u'(x)^2\, dx$$

    subject to

    $$u(0) \le 1, \quad u(1) = 1,$$

    $$\int_0^1 u(x)^2\, dx \le 2, \quad \int_0^1 u(x)\, dx = 1.$$

15. (An elementary obstacle problem) Try to understand the variational problem

    $$\text{Minimize} \quad \int_0^1 u'(x)^2\, dx$$

    subject to

    $$u(0) = u(1) = 0, \quad u(x) \ge u_0(x),$$

where $u_0(x)$ is a given function such that $u_0(0)$ and $u_0(1)$ are negative but u_0 is positive somewhere in the interval $(0,1)$. Find the optimal solutions when

$$u_0(x) = \frac{1}{8} - \frac{1}{2}x^2 + \frac{1}{2}x$$

and when

$$u_0(x) = \frac{1}{4}\sin(10x) - \frac{1}{2} - 4x^2 + 4x.$$

16. Exercise 12 of Chapter 1 can be understood and solved as a variational problem under pointwise constraints. Introduce a multiplier (a function) and try to determine the optimal solutions.

17. Find an approximate solution to the brachistochrone problem

$$I(u) = \int_0^1 \frac{\sqrt{1 + u'(x)^2}}{\sqrt{x}}\, dx,$$
$$u(0) = 0, \quad u(1) = -1,$$

as described in Section 5.9. Do the same for the hanging cable problem

$$\text{Minimize} \quad \int_0^1 u(x)\sqrt{1 + u'(x)^2}\, dx$$

subject to

$$u(0) = u(1) = 0, \quad \frac{3}{2} = \int_0^1 \sqrt{1 + u'(x)^2}\, dx.$$

Chapter 6

Optimal Control

1. INTRODUCTION

Optimal control is an important part of optimization, with many applications in different areas, especially in engineering. In this last chapter, we will simply study the basic ideas for tackling such problems. In particular, we will focus on Pontryagin's maximum principle, trying to insist on its importance through several examples.

The usual format of an optimal control problem is the following. The state of a certain system is described by a number of parameters,

$$x = (x_1, x_2, \ldots, x_n),$$

which evolve according to a state equation

$$x'(t) = f(t, x(t), u(t)),$$

where

$$u = (u_1, u_2, \ldots, u_m)$$

represents the control exercised on the system (with the objective in mind of controlling it). This control vector should typically satisfy various types of constraints depending on the nature of the problem. We will consider only the restriction $u(t) \in K \subset \mathbf{R}^m$ for all t, where K is prescribed a priori. The state equation is also completed with initial and/or final conditions

$$x(0) = x_0, \quad x(T) = x_T,$$

where T is the time horizon we are considering. We must also have a cost functional measuring how good a given control u is. The form of this objective functional is

$$I(x, u) = \int_0^T F(t, x(t), u(t)) \, dt,$$

where

$$F : (0, T) \times \mathbf{R}^n \times \mathbf{R}^m \to \mathbf{R}$$

is a known integrand associated with the cost we are willing to measure. A pair (x, u) is said to be feasible or admissible if
1. constraints on the control: $u(t) \in K$ for all $t \in (0, T)$;
2. state law: $x'(t) = f(t, x(t), u(t))$ for all $t \in (0, T)$;
3. end-point conditions: $x(0) = x_0$, $x(T) = x_T$.
As pointed out before, endpoint conditions may vary from having none of the two to having both. The optimal control problem consists in finding (determining or approximating) an admissible pair (X, U) such that

$$I(X, U) \leq I(x, u)$$

for all other feasible pairs (x, u).

Notice that this kind of problem includes, as a very particular case, variational problems in which we try to minimize

$$I(x) = \int_0^T F(t, x(t), x'(t)) \, dt$$

among all fields $x(t)$ with $x(0) = x_0$, $x(T) = x_T$. Indeed, this problem is equivalent to

$$x'(t) = u(t), \quad x(0) = x_0, \quad x(T) = x_T,$$

$$I(x, u) = \int_0^T F(t, x(t), u(t)) \, dt.$$

We will essentially focus in this chapter on necessary conditions of optimality. We are seeking conditions, in differential form, that optimal solutions to control problems should satisfy. We will take for granted that our problems are such that we always have the needed regularity to write down such optimality conditions. An entire fundamental field (nonsmooth optimization) deals with those situations in which this regularity cannot be taken for granted, and indeed, it is one of the main ingredients and main issues to be understood. Nonsmooth analysis is left for more advanced textbooks. See [9].

2. MULTIPLIERS AND THE HAMILTONIAN

We will first treat the case in which the set K is all of $\mathbf{R}^m$ so that there is no restriction on where admissible controls can take on values. Hence we are concerned with minimizing

$$I(x, u) = \int_0^T F(t, x(t), u(t)) \, dt$$

among all pairs (x, u) such that

$$x'(t) = f(t, x(t), u(t))$$

together with appropriate conditions on endpoints. As we have already noticed, the state equation may be considered as a pointwise constraint that can be treated by introducing a multiplier or costate $p(t)$. Therefore, we consider the augmented functional

$$I^*(x, u, p, x') = \int_0^T [F(t, x(t), u(t)) + p(t)(f(t, x(t), u(t)) - x'(t))] \, dt.$$

After our experience with constraints and multipliers, it seems plausible that optimal solutions for our initial optimal control problem ought to be a solution of the E-L equation for I^* regarded as a function of the four variables

(x, u, p, x'). Since Pontryagin's maximum principle is a more general statement than this and will be treated later, we will not insist on the validity of this assertion at this stage. If we put

$$G(t, u, p, x, u', p', x') = F(t, x, u) + p(f(t, x, u) - x'),$$

then the E-L system can be written

$$\frac{d}{dt}\left[\frac{\partial G}{\partial x'}\right] = \frac{\partial G}{\partial x}, \quad \frac{d}{dt}\left[\frac{\partial G}{\partial u'}\right] = \frac{\partial G}{\partial u}, \quad \frac{d}{dt}\left[\frac{\partial G}{\partial p'}\right] = \frac{\partial G}{\partial p},$$

that is,

$$0 = \frac{\partial F}{\partial x}(t, x, u) + p\frac{\partial f}{\partial x}(t, x, u) + p',$$

$$0 = \frac{\partial F}{\partial u}(x, u, t) + p\frac{\partial f}{\partial u}(t, x, u),$$

$$0 = x' - f(x, u, t).$$

By defining the Hamiltonian of the problem $H = F + pf$, these equations may be recast as

$$p' = -\frac{\partial H}{\partial x}, \quad \frac{\partial H}{\partial u} = 0, \quad x' = f(t, x, u).$$

This system must be completed with conditions on the endpoints. Notice that only two derivatives (x' and p') appear, so that in order to determine the solution we require two conditions. These are the endpoint restrictions for the state x completed with transversality conditions for the multiplier p according to the following rule:

Transversality statement 6.1 *If at a given endpoint (initial or final) we have a condition on the state, we do not enforce the corresponding transversality condition, but if the state is free, then the transversality condition $p = 0$ at the given endpoint must be taken into account.*

As we have already argued in many other situations, optimal solutions must be sought among the solutions of these optimality conditions. Let us look at several examples.

Example 6.2 *If*

$$I(x, u) = \int_0^1 u(t)^2 \, dt, \quad x'(t) = u(t) + ax(t),$$

$\mathsf{Inf}\ \{\ I(x,u)\ :\ x'(t) = u(t) + ax(t),\ x(0) = 1\}$

where $a \in \mathbf{R}$ is a constant, we would like to determine the optimal control under the initial condition $x(0) = 1$. In this example the Hamiltonian is

$$H = u^2 + p(u + ax),$$

(handwritten:) $\dfrac{\partial H}{\partial u} = 2u + p$　　$\dfrac{\partial H}{\partial x} = ap$

and the hypotheses that guarantee that optimal solutions are precisely the solutions to the optimality conditions are satisfied. This will be justified later. By manipulating these optimality conditions, we arrive at

$$u = -p/2, \quad p' = -ap, \quad x' = -p/2 + ax,$$

together with $x(0) = 1$, $p(1) = 0$. This last condition is the transversality condition at $t = 1$, since we do not have the final condition for the state (we will come back to this later). By solving for p we obtain

$$p(t) \equiv 0,$$

(handwritten:) $\dfrac{dp}{dt} = -ap$

and then

$$x(t) = e^{at}.$$

The optimal control is then $u \equiv 0$, as we could have anticipated. In reality, this situation does not have much interest, since there is, in fact, no demand on the system.

Assume now that we would like $x(0) = 1$ and $x(1) = 0$. The optimal strategy is no longer $u \equiv 0$, since this control leads the system to $x(1) = e^a$. Indeed, now we do not have any transversality condition for p, and

$$p(t) = ce^{-at}.$$

Taking this expression into the state equation, we obtain

$$x(t) = \frac{c}{4a}e^{-at} + de^{at}. \quad \text{?}$$

(handwritten:) Integrating factor

The constants c and d must be determined so that endpoint conditions are satisfied. This leads to a linear system whose solution is

$$c = \frac{2ae^a}{\sinh a}, \quad d = -\frac{e^{-a}}{2\sinh a}.$$

(handwritten:) $x(t) = \dfrac{p}{2a} + de^{at}$

(handwritten bottom:)

$x' - ax = -p/2$

$u(t) = e^{-\int a\, dt} = e^{-at}$

$e^{-at}(x' - ax) = -\dfrac{p}{2} \cdot e^{-at}$

$e^{-at} x(t) = -\dfrac{p}{2} \int e^{-at}\, dt = \dfrac{p}{2a} e^{-at} + d$

This time, the optimal control is

$$u(t) = -\frac{ae^{a(1-t)}}{\sinh a}.$$

Example 6.3   The cost is the same as in the previous example,

$$I(x, u) = \int_0^1 u(t)^2 \, dt,$$

while the equation of state is

$$x''(t) = u(t)$$

with endpoint conditions

$$x(0) = x'(0) = 1, \quad x(1) = 0.$$

The control u is the accelerator, and the cost is a measure of fuel expenditure. We first reduce the second order equation to a first order system with components $x_1 = x$, $x_2 = x'$, so that

$$x_1' = x_2, \quad x_2' = u,$$

and endpoint conditions

$$x_1(0) = 1, \quad x_1(1) = 0, \quad x_2(0) = 1.$$

The Hamiltonian is

$$H = u^2 + p_1 x_2 + p_2 u,$$

and optimality equations together with endpoint conditions are

$$\frac{\partial H}{\partial u} = 2u + p_2$$

$$\frac{\partial H}{\partial x_1} = 0$$

$$-\frac{\partial H}{\partial x_2} = p_1$$

$$u = -p_2/2,$$
$$p_1' = 0, \quad p_2' = -p_1,$$
$$x_1' = x_2, \quad x_2' = u,$$
$$x_1(0) = 1, \quad x_1(1) = 0,$$
$$x_2(0) = 1, \quad p_2(1) = 0.$$

It is not hard to find the optimal solution

$$u(t) = 6(t - 1),$$
$$x_1(t) = t^3 - 3t^2 + t + 1,$$
$$x_2(t) = 3t^2 - 6t + 1.$$

These optimality conditions are necessary for an optimal solution, but they might not be sufficient in the sense that there might be other types of solutions that are not optimal. At this point, the reader will not be surprised if we assert that convexity is involved in trying to establish that necessary conditions of optimality are also sufficient. We can, in fact, prove the following result. We stick to the initial notation for an optimal control problem

$$I(x, u) = \int_0^T F(t, x(t), u(t)) \, dt,$$
$$x'(t) = f(t, x(t), u(t)),$$
$$(x(0) = x_0), \quad (x(T) = x_T),$$

We have placed endpoint conditions within parentheses to indicate that they may be present or not.

Theorem 6.4   *Let f be linear in (x, u) and F convex in (x, u) for each fixed t. Then every solution of the system of optimality with the appropriate endpoint conditions (including transversality) will be an optimal solution of the control problem.*

Imagine that the pair (x, u) satisfies all the optimality conditions, and let $(\tilde{x}, \tilde{u})$ be any other admissible pair. We will measure the difference

$$I(\tilde{x}, \tilde{u}) - I(x, u)$$

and conclude that it cannot be negative. This implies that (x, u) is indeed optimal.

Due to the hypotheses of linearity and convexity assumed in the statement of the theorem, we can write

$$I(\tilde{x}, \tilde{u}) - I(x, u) = \int_0^T [F(t, \tilde{x}, \tilde{u}) - F(t, x, u)]\, dt$$

$$\geq \int_0^T \left[\frac{\partial F}{\partial x}(t, x, u)(\tilde{x} - x) + \frac{\partial F}{\partial u}(t, x, u)(\tilde{u} - u) \right] dt$$

$$= \int_0^T \left[\left(-p\frac{\partial f}{\partial x}(t, x, u) - p'\right)(\tilde{x} - x) - p\frac{\partial f}{\partial u}(t, x, u)(\tilde{u} - u) \right] dt$$

$$= -\int_0^T p\left[(x' - \tilde{x}') + \frac{\partial f}{\partial x}(t, x, u)(\tilde{x} - x) + \frac{\partial f}{\partial u}(t, x, u)(\tilde{u} - u) \right] dt$$

$$= -\int_0^T p\left[f(t, x, u) - f(t, \tilde{x}, \tilde{u}) + \frac{\partial f}{\partial x}(t, x, u)(\tilde{x} - x) + \frac{\partial f}{\partial u}(t, x, u)(\tilde{u} - u) \right] dt$$

$$= 0.$$

Note how endpoint conditions and transversality are used to check that contributions coming from endpoints vanish in the integrations by parts.

Sometimes, we may have to enforce additional integral constraints.

Example 6.5 *Assume a particular process described by a function $x(t)$ starting at $x(0) = 1$. Our desire is to lead the system to $x(T) = 0$ in the least time possible, where the system evolves according to the law*

$$x'(t) = ax(t) + u(t),$$

where a is a given constant, and we must respect the constraint

$$K = \int_0^T u(t)^2\, dt,$$

for some fixed given constant $K > 0$. It is precisely this integral constraint that makes the problem interesting, because otherwise, T could be as small as we like. After our experience with integral constraints in the last chapter, we are going to introduce a multiplier $s \in \mathbf{R}$ associated with this restriction, and put

$$H = 1 + su^2 + p(ax + u)$$

for the Hamiltonian, where s is to be determined later. In fact, notice that we can also write

$$H = s \left(\frac{1}{s} + u^2 + \frac{p}{s}(ax + u) \right),$$

and replacing p by p/s, we realize that optimality conditions are the same for the Hamiltonian

$$H = 1 + u^2 + p(ax + u),$$

but in this case, there is no additional constant s. It has been incorporated somehow in the costate p. The optimality equations are

$$p' = -ap, \quad 2u + p = 0, \quad x' = ax + u.$$

By solving for u and p and replacing them in the state equation, we obtain

$$x' = ax + ce^{-at}$$

for a certain constant c. By solving this last equation, we arrive at

$$x(t) = de^{at} + \frac{c}{2a}e^{-at}.$$

We impose the three conditions

$$x(0) = 1, \quad x(T) = 0, \quad K = \int_0^T u(t)^2 \, dt,$$

and have

$$1 = d + \frac{c}{2a},$$

$$0 = de^{aT} + \frac{c}{2a}e^{-aT},$$

$$K = \frac{c^2}{8a}\left(1 - e^{-2aT}\right).$$

By manipulating these equations we get the minimal T,

$$T = -\frac{1}{2a}\log\left(1 - \frac{a}{2K}\right),$$

the optimal control

$$u(t) = -2Ke^{-at},$$

and the associated optimal state

$$x(t) = e^{at} - \frac{4K}{a}\sinh(at).$$

Notice that the formula for T requires $a < 2K$. If $a = 2K$, then $T = +\infty$, and the system would tend to 0 as $t \to \infty$ but never reach it. If $a > 2K$, the system does not even approach the state 0, since $x(t)$ tends to $+\infty$.

All the examples we have looked at so far lie within the hypotheses of Theorem 6.4, so that the solutions we have found are indeed the optimal solutions.

3. PONTRYAGIN'S PRINCIPLE

If we want to stay closer to more realistic hypotheses for optimal control problems, we must care about the possibility of having some bounds on the size of admissible controls. This is reasonable, since a priori, a given system may not be able to withstand the effect of a control of arbitrary size, either because it would collapse, or because it would get out of the regime in which the state law is valid or simply because we do not have the means to apply a control of arbitrary size. In any case, a restriction on the size of feasible controls must be considered. Typically, this is formulated by requiring

$$u(t) \in K,$$

where K is an appropriate subset of the space in which controls take on values. Hence, in this section we are concerned with the problem

$$\text{Minimize} \quad \int_0^T F(t, x(t), u(t))\, dt$$

subject to

$$x'(t) = f(t, x(t), u(t)),$$
$$u(t) \in K, \quad x(0) = x_0, \quad (x(T) = x_T).$$

We are especially interested in understanding necessary conditions of optimality that optimal solutions should satisfy.

If we recall the discussion relative to Bellman's equation in the last chapter, we may consider the value function

$$S(t, x) = \min_u \left\{ \int_t^T F(\tau, y(\tau), u(\tau)) \, d\tau : y'(\tau) = f(\tau, y(\tau), u(\tau)), \right.$$

$$\left. u(\tau) \in K, y(t) = x \right\}.$$

The fundamental property that the value function must satisfy for $t' > t$ is

$$S(t, x) = \min_y \left\{ \min_v \left\{ \int_t^{t'} F(\tau, z(\tau), v(\tau)) \, d\tau : z'(\tau) = f(\tau, z(\tau), v(\tau)), v(\tau) \in K, \right. \right.$$

$$\left. \left. z(t) = x, z(t') = x + y(t' - t) \right\} + S(t', x + y(t' - t)) \right\}.$$

This condition can be rewritten as

$$0 = \min_y \left\{ \min_v \left\{ \frac{1}{t' - t} \int_t^{t'} F(\tau, z(\tau), v(\tau)) \, d\tau : z'(\tau) = f(\tau, z(\tau), v(\tau)), \right. \right.$$

$$\left. v(\tau) \in K, z(t) = x, z(t') = x + y(t' - t) \right\}$$

$$\left. + \frac{S(t', x + y(t' - t)) - S(t, x)}{t' - t} \right\}.$$

Taking limits as $t' \searrow t$, we conclude (why?) that

$$0 = \min_y \left\{ \min_{v \in K} \{ F(t, x, v) : y = f(t, x, v) \} + \frac{\partial S}{\partial t}(t, x) + y \frac{\partial S}{\partial x}(t, x) \right\},$$

which can also be reorganized as

$$\frac{\partial S}{\partial t}(t, x) = - \min_{v \in K} \left[F(t, x, v) + f(t, x, v) \frac{\partial S}{\partial x}(t, x) \right]. \tag{6-1}$$

The question now is, what kind of information does this equation provide concerning an optimal pair $(x(t), u(t))$? The fact that $(x(t), u(t))$ is optimal means that

$$S(t, x(t)) = \int_t^T F(\tau, x(\tau), u(\tau)) \, d\tau, \quad x'(\tau) = f(\tau, x(\tau), u(\tau)),$$

for every time t. If we differentiate with respect to t and use (6–1), we can write

$$-F(t, x(t), u(t)) = \frac{\partial S}{\partial t}(t, x(t)) + x'(t)\frac{\partial S}{\partial x}(t, x(t))$$

$$= -\min_{v \in K}\left[F(t, x(t), v) + f(t, x(t), v)\frac{\partial S}{\partial x}(t, x(t))\right]$$

$$+ f(t, x(t), u(t))\frac{\partial S}{\partial x}(t, x(t)),$$

and hence

$$F(t, x(t), u(t)) + f(t, x(t), u(t))\frac{\partial S}{\partial x}(t, x(t))$$

$$= \min_{v \in K}\left[F(t, x(t), v) + f(t, x(t), v)\frac{\partial S}{\partial x}(t, x(t))\right].$$

On the other hand, if we put $p(t, x) = \frac{\partial S}{\partial x}(t, x)$, on the optimal pair we should have, by using again

$$\frac{\partial S}{\partial t}(t, x(t)) = x'(t)\frac{\partial S}{\partial x}(t, x(t)) + F(t, x(t), u(t)),$$

that

$$\frac{d}{dt}p(t, x(t)) = \frac{\partial^2 S}{\partial x \partial t}(t, x(t)) + f(t, x(t), u(t))\frac{\partial^2 S}{\partial x^2}(t, x(t))$$

$$= -\frac{\partial}{\partial x}\left(F(t, x(t), u(t)) + f(t, x(t), u(t))\frac{\partial S}{\partial x}(t, x(t))\right)$$

$$+ f(t, x(t), u(t))\frac{\partial^2 S}{\partial x^2}(t, x(t))$$

$$= -\frac{\partial F}{\partial x}(t, x(t), u(t)) - p(t, x(t))\frac{\partial f}{\partial x}(t, x(t), u(t)).$$

By means of the Hamiltonian

$$H(t, x, u, p) = F(t, x, u) + pf(t, x, u)$$

and the function $p(t) = p(t, x(t))$, we can summarize the above conclusions as in the next statement. Notice the relationship between the multiplier (costate) $p(t)$ and the value function $S(t, x)$:

$$p(t) = p(t, x(t)) = \frac{\partial S}{\partial x}(t, x(t))$$

if $x(t)$ is optimal.

Theorem 6.6 (Pontryagin's principle: necessary conditions). *If the pair* $(x(t), u(t))$ *is optimal for our original control problem, there must exist a function* $p(t)$ *such that the following conditions hold:*

$$p'(t) = -\frac{\partial H}{\partial x}(t, x(t), u(t), p(t)), \quad (p(T) = 0),$$

$$H(t, x(t), u(t), p(t)) = \min_{v \in K} H(t, x(t), v, p(t)),$$

$$x'(t) = f(t, x(t), u(t)), \quad x(0) = x_0, \quad (x(T) = x_T).$$

Notice how the second condition on the above statement is formulated as an NLPP for v depending on various parameters. We have written the transversality condition and the final condition on the state within parentheses to emphasize that one of the two, but not both at the same time, must be enforced. The only comment on this important statement concerns precisely the transversality condition $p(T) = 0$. If we keep in mind the definition of $p(t)$ as

$$p(t) = \frac{\partial S}{\partial x}(t, x(t)),$$

the transversality condition means that

$$\frac{\partial S}{\partial x}(T, x(T)) = 0.$$

This constraint reflects nothing but the fact that if the value at the right endpoint $x(T)$ is free, the value function must attain its minimum when it takes on the value given by the optimal state $x(T)$. Consequently, the derivative should vanish. If in the original formulation of the control problem we have a constraint on the final state so that it is fixed, $x(T) = x_T$, then this condition replaces transversality. The same discussion applies at the initial time $t = 0$.

It is also interesting to realize that the conditions on the above result are a generalization of those in Section 6.2 since if K is all of $\mathbf{R}^m$, then the minimum of the Hamiltonian on v is achieved when the derivative with respect to v vanishes.

Before proceeding to study sufficient conditions of optimality, we look at several examples to examine Pontryagin's principle.

One of the most common families of optimal control problems is that concerned with performing a given task in the least time possible. In such cases, the objective functional to be minimized is the time employed in competing the given task.

Example 6.7   *Imagine, to begin with, a mobile object whose movement we can control with its accelerator u, where the maximum allowable acceleration is b, and maximum brake power is $-a$; i.e., $-a \leq u \leq b$. According to our previous setting, $K = [-a, b]$. Starting out at rest and ending up at rest, we would like to travel a distance α in minimum time. What is the optimal strategy for using the accelerator? The cost functional is*

$$I = \int_0^T 1 \, dt,$$

under the restrictions

$$x''(t) = u(t), \quad u \in K,$$
$$x(0) = x'(0) = 0, \quad x(T) = \alpha, \quad x'(T) = 0.$$

If we transform this second order equation into a first order system in the standard way by means of the change

$$x_1(t) = x(t), \quad x_2(t) = x'(t),$$

we obtain

$$x_1' = x_2, x_2' = u, \quad u \in K,$$
$$x_1(0) = x_2(0) = 0, \quad x_1(T) = \alpha, \quad x_2(T) = 0.$$

The Hamiltonian of the system will be

$$H(t, x, u, p) = 1 + p_1 x_2 + p_2 u,$$

and the necessary conditions of optimality involving p, according to Pontryagin's principle, will be

$$p_1' = 0, \quad p_2' = -p_1.$$

Hence $p_1 = -d$ is constant, and $p_2 = dt + c$, with c, d constants. If we take this information back to the condition on the minimum, we obtain that the optimal control $u(t)$ should minimize the Hamiltonian H over K. In our situation,

$$(dt + c)u(t) = \min_{-a \leq v \leq b}(dt + c)v.$$

$$\frac{\partial H}{\partial u} = p_2$$

$$\Rightarrow u = p_2$$

Since the expression $(dt + c)v$ is linear in v, the previous minimum is attained at one of the endpoints of the interval $[-a, b]$ depending on the sign of $(dt + c)$. Therefore, the optimal control will have the form

$$u(t) = \begin{cases} -a, & dt + c > 0, \\ b, & dt + c < 0, \\ \text{any value}, & dt + c = 0. \end{cases}$$

Due to the physical interpretation of our problem, we have $u(t) = b$ at the beginning, since otherwise, our vehicle will not move, and later, obviously, we will have to use the brake. Notice also that the above form for the optimal control discards the possibility of having several changes between positive and negative accelerations, since $dt + c$, being linear in t, can pass through zero at most once. Consequently, the optimal control will have the form

$$u(t) = \begin{cases} b, & t \leq t_0, \\ -a, & t \geq t_0. \end{cases}$$

The instant t_0 must be determined in terms of a, b, and α. Indeed, we will have to solve

$$x''(t) = b, \quad x(0) = x'(0) = 0,$$

with solution

$$x(t) = bt^2/2.$$

At time t_0 (to be determined) a change in the dynamics of the system takes place, and we must solve

$$x''(t) = -a, \quad x(t_0) = bt_0^2/2, \quad x'(t_0) = bt_0,$$

asking, moreover, for $x(T) = \alpha$ and $x'(T) = 0$. Thus

$$x(t) = bt_0 \left(t - \frac{t_0}{2} \right) - a\frac{(t - t_0)^2}{2},$$

and the conditions at time T lead to a system of two equations in the two unknowns T and t_0. After some computations we have

$$t_0 = \sqrt{\frac{2a\alpha}{b(a + b)}}, \quad T = \sqrt{\frac{2(a + b)\alpha}{ab}}.$$

Example 6.8   *Our next example is also a minimum time problem, so that the objective functional is the same, but the state system is now*

$$x'_1 = -x_1 + u, x'_2 = u, \quad |u| \leq 1,$$

under arbitrary initial conditions

$$x_1(0) = a, \quad x_2(0) = b.$$

The task consists in leading the system to rest,

$$x_1(T) = x_2(T) = 0$$

in minimum time.

The Hamiltonian is

$$H(t, x, u, p) = 1 + p_1(-x_1 + u) + p_2 u,$$

which, as in the preceding example, is linear in u. The equations for p_1 and p_2 are

$u = P_1 + P_2$ $$p'_1 = p_1, \quad p'_2 = 0,$$

so that $p_2 = d$ and $p_1 = ce^t$, where c and d are constants. Due to the linearity of H with respect to u, the optimal control will have the form

$$u(t) = \begin{cases} -1, & p_1 + p_2 > 0, \\ 1, & p_1 + p_2 < 0, \\ \text{any value}, & p_1 + p_2 = 0. \end{cases}$$

We must ask ourselves how many times the expression

$$p_1 + p_2 = d + ce^t$$

can go through the origin, with the aim of deciding how many changes from −1 to 1 the optimal control will exhibit, and then, through the appropriate computations, find the optimal strategy. Since the initial conditions are arbitrary and not given explicitly, the discussion in terms of the specific formulas providing the optimal time and the instants where a change in the optimal control should take place in terms of a and b becomes really tedious and almost impossible to clarify. On such occasions, it is much more fruitful to analyze the problem in a

qualitative fashion by means of "switching curves." We describe in the sequel
this sort of analysis for the example at hand.

As we have argued above, the optimal control makes the process alternate
between the dynamics of the two systems

$$\begin{cases} x_1' = -x_1 + 1, \\ x_2' = 1, \end{cases} \qquad \begin{cases} x_1' = -x_1 - 1, \\ x_2' = -1. \end{cases}$$

The integral curves of these systems starting at (a, b) when $t = 0$ are, respectively,

$$x_1(t) = (a - 1)e^{-t} + 1, \quad x_2(t) = t + b,$$

and

$$x_1(t) = (a + 1)e^{-t} - 1, \quad x_2(t) = -t + b.$$

They are schematically drawn in Figure 6.1, where the integral curve through
the origin has been highlighted in each case.

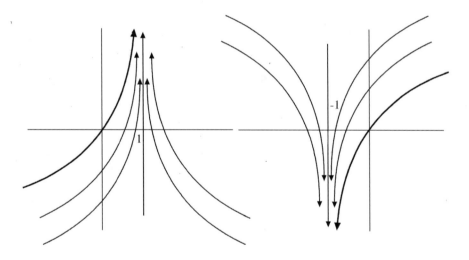

Figure 6.1. Integral curves of a control system.

If we imagine these two families of curves in a single diagram, the choice of
the optimal control is dictated by starting at the given point (a, b), following

one of the corresponding integral curves, and, at a certain instant, changing to the other one in such a way that the second integral curve must take us to the origin, if this is possible. Since only one integral curve of each system will pass through the origin (in fact, since time is not allowed to decrease, we are talking about two half-curves), it is a matter of reaching one of these two half-curves as soon as possible starting from the given initial point. Specifically, these two half-curves are

$$\Lambda_1 = \left\{ (1 - e^{-s}, s) : s \leq 0 \right\} \quad \text{for } u = 1,$$
$$\Lambda_{-1} = \left\{ (e^s - 1, s) : s \geq 0 \right\} \quad \text{for } u = -1.$$

Let Λ be the union of these two curves, and let Λ^+ designate the part of the plane over Λ, and Λ^- the part of the plane under Λ. It is easy to see that if the point (a, b) belongs to Λ^+, we should take $u = -1$ until we are on Λ_1, changing in this instant to $u = 1$, since that curve will lead us to the origin. In the same way, for initial conditions $(a, b) \in \Lambda^-$, we must choose first $u = 1$ until reaching Λ_{-1}, and then $u = -1$ since Λ_{-1} will lead us to rest. These are the optimal strategies, since any other way of changing between these two dynamics will make us "waste some time." See Figure 6.2.

It is also interesting to point out that the number of changes in the optimal control can also be determined by going back to the equation

$$p_1 + p_2 = d + ce^t.$$

The number of changes corresponds to the roots of the equation

$$d + ce^t = 0,$$

for arbitrary constants c, d. By solving for t, we obtain

$$t = \log\left(-\frac{d}{c}\right).$$

We thus conclude that the optimal control will have at most (when d/c is negative) one switch.

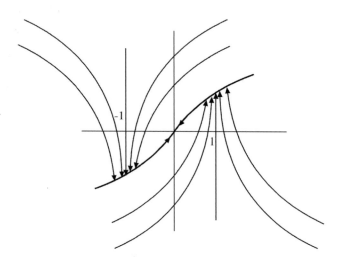

Figure 6.2. Draft of optimal strategies.

Example 6.9   One interesting problem that can be understood by a qualitative analysis using the switching-curve concept is the optimal control of a harmonic oscillator, which is described by the equations

$$x_1' = x_2, \quad x_2' = -x_1 + u,$$

where the control u is restricted in size: $|u| \le 1$. Again, we would like to know how to lead the system to rest starting from an arbitrary initial state (a_1, a_2).

The Hamiltonian is

$$H(t, x, u, p) = 1 + p_1 x_2 + p_2(-x_1 + u).$$

Since it is linear in u, optimal strategies will always alternate between $u = 1$ and $u = -1$. Those two dynamics are represented by the integral curves of the systems

$$\begin{cases} x_1' = x_2, \\ x_2' = -x_1 + 1, \end{cases} \qquad \begin{cases} x_1' = x_2, \\ x_2' = -x_1 - 1. \end{cases}$$

It is not hard to check that the integral curves for the first are concentric circles centered at $(1, 0)$, while those for the second are concentric circles centered at $(-1, 0)$ (Figure 6.3).

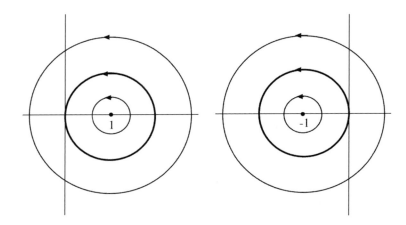

Figure 6.3. Integral curves for the harmonic oscillator.

The issue is to understand optimal strategies for arbitrary initial points in the plane. To this end, we examine the differential equations for the costates and see what information we can derive from them. Those are

$$p_1' = p_2,$$
$$p_2' = -p_1.$$

By differentiation we have

$$p_2'' + p_2 = 0,$$

so that

$$p_2(t) = A\cos(t + B)$$

for arbitrary constants A, B. We know that the changes in the optimal control are dictated by the times that the costate p_2 passes through the origin.

Due to the form of p_2, we conclude that there might be an arbitrary number of changes (depending on the initial conditions) and that these are to be produced before π units of time goes by. Notice that roots of $\cos(B + t)$ are located π units apart from each other. By bearing in mind this information, and examining first those points at which no switch in the control is necessary, then those requiring a single change, two changes, and so on, it is possible to understand optimal strategies. These conclusions are indicated in Figure 6.4.

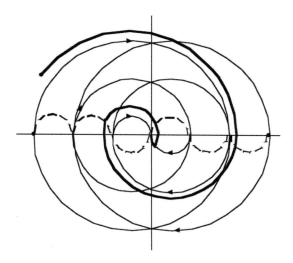

Figure 6.4. Some optimal paths for the harmonic oscillator.

Example 6.10   From a given fixed point, a projectile is launched with the aim of hitting a target located $3 + 5/6$ distance units away in 3 time units. We want this goal to be accomplished with the least cost measured by the integral

$$E = k \int_0^3 u(t)^2 \, dt, \quad k > 0,$$

where the control u indicates the accelerator of the projectile. The state equation is $x'' = u$, and u is restricted in sign and size: $0 \le u \le 1$. The initial and final conditions are then

$$x'(0) = x(0) = 0, \quad x(3) = 3 + \frac{5}{6}.$$

Transforming the second order equation to a first order system, we immediately obtain the Hamiltonian

$$H(t, x, u, p) = ku^2 + p_1 x_2 + p_2 u,$$

which is a strictly convex function of u. The optimality conditions state that

$$p_1' = 0, \quad p_2' = -p_1,$$

$$p_1(t) = C$$
$$p_2(t) = Ct + d$$

together with the transversality condition $p_2(3) = 0$, since the value $x_2(3)$ is free. Consequently, we obtain $p_2(t) = d(3 - t)$, where $p_1 = d$ is constant.

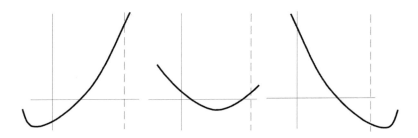

Figure 6.5. Quadratic dependence of the Hamiltonian on the control.

The condition on the minimum of H over $u \in [0, 1]$, which is quadratic, leads to three situations, depending on whether the vertex of the parabola lies in $[0, 1]$ or stays to the right or left of this interval. These three cases are (see Figure 6.5):

1. $u = -p_2/2k = c(t - 3)$, $c = -d/3k$ if $c(t - 3) \in (0, 1)$;
2. $u = 0$ if $c(t - 3) \leq 0$;
3. $u = 1$ if $c(t - 3) \geq 1$.

Nonetheless, since the control cannot be identically zero (because the projectile would not move) only two cases are possible, in which c is always nonpositive (Figure 6.6):

1. $0 < c(t - 3) < 1$, $0 \leq t < 3$: In this case, we should solve the problem

$$x'' = c(t - 3), \quad x(0) = x'(0) = 0, \quad x(3) = 3 + \frac{5}{6}.$$

After a few computations, we obtain

$$x(t) = \frac{c}{6}t^3 - \frac{3c}{2}t^2, \quad c = -\frac{1}{3} - \frac{5}{54}.$$

However, for this value of c the optimal control $u(t) = c(t - 3)$ always violates the restriction $u \in [0, 1]$, since $u(0) > 1$. This cannot be the solution sought.

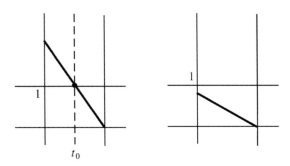

Figure 6.6. Possibilities for $c(t - 3)$.

2. *There exists $t_0 \in (0, 1)$ such that $c(t_0 - 3) = 1$. In this situation*

$$u(t) = \begin{cases} 1, & t \leq t_0, \\ c(t - 3), & t \geq t_0. \end{cases}$$

We solve the problem

$$x'' = u, \quad x(0) = x'(0) = 0, \quad x(3) = 3 + \frac{5}{6},$$

in two steps. Firstly,

$$x''(t) = 1 \text{ for } t \leq t_0, \quad x(0) = x'(0) = 0,$$

has the solution $x(t) = t^2/2$. Next, imposing the continuity at t_0 of x and x',

$$x''(t) = c(t - 3) \quad \text{for} \quad t \geq t_0,$$

$$x(t_0) = \frac{t_0^2}{2}, \quad x'(t_0) = t_0, \quad x(3) = 3 + \frac{5}{6},$$

where we bear in mind that $c(t_0 - 3) = 1$. After performing all these substitutions, we are led to a polynomial equation for t_0,

$$t_0^3 - 9t_0^2 + 23t_0 - 15 = 0, \quad t_0 \in (0, 3).$$

The only admissible value is $t_0 = 1$, and therefore, $c = -1/2$. The optimal control is

$$u(t) = \begin{cases} 1, & t \leq 1, \\ -\frac{t}{2} + \frac{3}{2}; & t \geq 1. \end{cases}$$

Example 6.11   *Our next example describes a situation in which the control has two components to highlight the main differences with respect to the examples examined above.*

Suppose a system obeys the state equation

$$x_1' = -x_2 + u_1, \quad x_2' = x_1 + u_2,$$

where this time the control we exercise on the system should respect the contraint

$$u_1^2 + u_2^2 \leq 1.$$

We would like to determine the shortest time interval in which we can lead the system to rest from an arbitrary initial condition (a, b). The Hamiltonian is

$$H(t, x, u, p) = 1 + p_1(-x_2 + u_1) + p_2(x_1 + u_2),$$

and the condition on the minimum over the control (u_1, u_2) reads

$$p_1 u_1 + p_2 u_2 = \min\left\{p_1 v_1 + p_2 v_2 : v_1^2 + v_2^2 \leq 1\right\}.$$

This is a simple exercise of nonlinear programming (Chapter 3). At this point, our readers should not have any difficulty in finding the optimal solution

$$u_1 = -\frac{p_1}{\sqrt{p_1^2 + p_2^2}}, \quad u_2 = -\frac{p_2}{\sqrt{p_1^2 + p_2^2}},$$

which provide the optimal control once multipliers (costates) are known. The equations for those are

$$p_1' = -p_2, \quad p_2 = p_1.$$

Plugging one of them into the other, we arrive at

$$p_1(t) = \rho_0 \cos(t + \theta_0), \quad p_2(t) = \rho_0 \sin(t + \theta_0),$$

where we have used the form $\rho_0 \cos(t + \theta_0)$ with ρ_0, θ_0 arbitrary constants, for the general solution of the equation $p_1'' + p_1 = 0$. In this way, the optimal control is

$$u_1(t) = -\cos(t + \theta_0), \quad u_2(t) = -\sin(t + \theta_0).$$

We now use the state system

$$x_1' = -x_2 - \cos(t + \theta_0), \quad x_2' = x_1 - \sin(t + \theta_0).$$

Once again, by differentiating and plugging one into the other, we obtain

$$x_1'' + x_1 = 2\sin(t + \theta_0).$$

The general solution of this equation is

$$x_1(t) = -t\cos(t + \theta_0) + \rho_1\cos(t + \theta_1),$$

and consequently,

$$x_2(t) = -t\sin(t + \theta_0) + \rho_1\sin(t + \theta_1).$$

Finally, initial and final conditions lead to

$$a = \rho_1\cos\theta_1, \quad b = \rho_1\sin\theta_1,$$
$$0 = -T\cos(T + \theta_0) + \rho_1\cos(T + \theta_1),$$
$$0 = -T\sin(T + \theta_0) + \rho_1\sin(T + \theta_1).$$

From this, it is immediate to obtain

$$T = \rho_1 = \sqrt{a^2 + b^2}, \quad \theta_0 = \theta_1,$$

and the optimal strategy

$$u_1(t) = -\cos(t + \theta_0), \quad u_1(t) = -\sin(t + \theta_0),$$

where θ_0 is the argument of the initial state (a, b).

One of the most important differences we discover in this example, in contrast to situations in which the control is a single number, is that the boundary of a connected region in two or more variables is not disconnected, and thus optimal controls do not need to jump abruptly (though sometimes do) from one point to another, but changes in the dynamics of the system may take place smoothly.

We have just studied a number of examples in which, by means of Pontryagin's principle, we have apparently found optimal solutions to problems. As in other situations analyzed before, the question is whether we can convince ourselves that those computed optimal solutions are indeed optimal. This is obviously an important issue. It is the matter of whether necessary conditions of optimality are sufficient. Once again, convexity enters in a relevant way into the discussion.

Theorem 6.12 (Sufficiency of optimality conditions)   *Assume that $F(t, x, u)$ is convex in (x, u), and $f(t, x, u)$ linear in (x, u). If the triplet $(x(t), u(t), p(t))$ satisfies*

$$p'(t) = -\frac{\partial H}{\partial x}(t, x(t), u(t), p(t)), \quad (p(T) = 0),$$

$$H(t, x(t), u(t), p(t)) = \min_{v \in K} H(t, x(t), v, p(t)),$$

$$x'(t) = f(t, x(t), u(t)), \quad x(0) = x_0, \quad (x(T) = x_T),$$

where

$$H(t, x, u, p) = F(t, x, u) + pf(t, x, u)$$

is the Hamiltonian of the system, and K is a convex subset, then the pair $(x(t), u(t))$ is an optimal solution of the corresponding optimal control problem.

This sufficiency result is not difficult to justify after the experience we already have with convexity. The condition on the minimum means that

$$g(s) = H(t, x(t), u(t) + s(v - u(t)), p(t)), \quad s \in [0, 1],$$

as a function of s, for fixed t and $v \in K$, has a minimum at $s = 0$. Notice how the vector $sv + (1 - s)u(t)$ belongs to K if this set is convex. In this situation all we can ensure is $g'(0) \geq 0$ (one-sided minimum). Therefore,

$$0 \leq g'(0) = \frac{\partial H}{\partial u}(t, x(t), u(t), p(t))(v - u(t)), \quad v \in K. \tag{6-2}$$

If $(\tilde{x}(t), \tilde{u}(t))$ is any other feasible pair for our control problem, we have

$$I(\tilde{x}, \tilde{u}) - I(x, u)$$
$$= \int_0^T [F(t, \tilde{x}(t), \tilde{u}(t)) + p(t)(f(t, \tilde{x}(t), \tilde{u}(t)) - \tilde{x}'(t))$$
$$\qquad -F(t, x(t), u(t)) - p(t)(f(t, x(t), u(t)) - x'(t))] \, dt$$
$$= \int_0^T [H(t, \tilde{x}(t), \tilde{u}(t), p(t)) - H(t, x(t), u(t), p(t))$$
$$\qquad -p(t)(\tilde{x}'(t) - x'(t))] \, dt$$
$$\geq \int_0^T \left[\frac{\partial H}{\partial x}(t, x(t), u(t), p(t))(\tilde{x}(t) - x(t)) \right.$$
$$\qquad \left. + \frac{\partial H}{\partial u}(t, x(t), u(t), p(t))(\tilde{u} - u(t)) + p'(t)(\tilde{x}(t) - x(t)) \right] \, dt$$
$$\geq 0,$$

due to the equation that $p(t)$ satisfies and (6–2). Hence the pair $(x(t), u(t))$ is truly an optimal solution. We have also utilized in an essential way the convexity of H with respect to (x, u), which is guaranteed because of the convexity of F and the linearity of f. The presence of the multiplier in front of f (and in particular its sign) keeps us from apparently relaxing the linearity of f. This issue is left to a more advanced course in optimal control.

Quite often, optimal solutions for control problems cannot be found, either because there are many (or not so many) variables involved, so that it is almost impossible to handle them all by hand; or else because optimality conditions cannot be solved explicitly or it is really cumbersome and tedious to find explicit formulas.

Example 6.13   *A mobile object in a plane can be controlled by two parameters, r_1 and r_2, expressing the rapidity with which the direction of movement can be changed (angular velocity of movement) and the modulus of velocity, respectively. The equations of motion are*

$$x''(t) = \cos\theta(t)r_2(t), \quad y''(t) = \sin\theta(t)r_2(t), \quad \theta'(t) = r_1(t).$$

Restrictions on the feasible pairs (r_1, r_2) can be generally written by requiring

$$(r_1, r_2) \in K,$$

where K is the set of admissible controls. The objective is to change the position of the object from, say, (a_0, b_0) to (a_1, b_1) in minimum time. The equivalent first order system is

$$x_1' = x_2, \quad x_2' = \cos\theta r_2,$$
$$x_3' = x_4, \quad x_4' = \sin\theta r_2,$$
$$\theta' = r_1,$$

and the Hamiltonian is

$$H = 1 + p_1 x_2 + p_2 \cos\theta r_2 + p_3 x_4 + p_4 \sin\theta r_2 + p_5 r_1.$$

The conditions of optimality are written

$$p_1' = 0, \quad p_2' = -p_1,$$
$$p_3' = 0, \quad p_4' = -p_3,$$
$$p_5' = p_2 r_2 \sin\theta - p_4 r_2 \cos\theta,$$
$$x'' = \cos\theta r_2,$$
$$y'' = \sin\theta r_2, \quad \theta' = r_1,$$

where $r = (r_1, r_2)$ must, in turn, be the optimal solution of

$$\min_{(r_1, r_2) \in K} \left((p_2 \cos\theta + p_4 \sin\theta)r_2 + p_5 r_1 \right).$$

Even for simple choices of the set K (as a rectangle or an ellipse), it is almost impossible to fully determine in a explicit form the optimal solution. On the other hand, in this particular situation restrictions on the state (in addition to those on the controls) in the form of obstacles to be avoided are very natural. This, however, is well beyond the scope of this text.

Example 6.14   The economy of a certain country follows the law

$$k' = f(k) - (\lambda + \mu)k - c,$$

where k is the ratio of invesment per unit of labor, f is the production function, μ and λ are parameters related to the depreciation and labor growth respectively, and c is consumption per unit of labor. The objective of this country is to choose consumption so as to maximize the welfare integral over a fixed time interval

$$\int_0^T e^{-\delta t} u(t)\, dt$$

where δ is the time discount parameter and u is the utility function. This last function must satisfy the equation

$$\eta = -\frac{cu''}{u'}$$

for a constant η, which is called the elasticity of marginal utility. If we assume that k is known both at the beginning and the final time and $u(T)$ is also known, we would like to determine the optimal consumption.

If we change the notation to make the formulation more transparent, we find, by putting

$$x_1 = k, \quad x_2 = u, \quad x_3 = u', \quad v = c,$$

that the Hamiltonian is

$$H = e^{-\delta t} x_2 + p_1(f(x_1) - (\lambda + \mu)x_1 - v) + p_2 x_3 - p_3 \eta \frac{x_3}{v},$$

and the optimality conditions read

$$x_1' = f(x_1) - (\lambda + \mu)x_1 - v,$$
$$x_2' = x_3,$$
$$x_3' = -\eta \frac{x_3}{v},$$
$$p_1' = -f'(x_1) + \lambda + \mu,$$
$$p_2' = e^{-\delta t},$$
$$p_3' = -p_2 + \frac{p_3 \eta}{v},$$
$$v^2 = \frac{p_3 \eta x_3}{p_1}.$$

Notice how the dependence of H on v is convex when $v > 0$. This system of six coupled differential equations is completed with endpoint conditions and transversality conditions, namely,

$$x_1(0) = k_0, \quad x_1(T) = k_T, \quad x_2(T) = u_T,$$
$$p_2(0) = p_3(0) = p_3(T) = 0.$$

The whole system is impossible to solve explicitly.

It is not especially hard to check in all the examples examined above that the hypotheses of the sufficiency optimality criterion are satisfied, so that solutions found are truly optimal solutions in all cases.

When these sufficiency conditions ensuring optimality are violated, then nonexistence of optimal solutions may follow. One of the simplest such examples is concerned with minimizing

$$\int_0^1 \left[(u(t)^2 - 1)^2 + x(t)^2 \right] dt,$$

where $x'(t) = u(t)$, $x(0) = 0$ and $K = \mathbf{R}$. If we take

$$u_j(t) = 1, \quad t \in \left(\frac{k}{2^j}, \frac{k+1}{2^j} \right), \quad k \text{ even},$$

$$u_j(t) = -1, \quad t \in \left(\frac{k+1}{2^j}, \frac{k+2}{2^j} \right), \quad k \text{ odd},$$

it is easy to check that $I(u_j) \searrow 0$, and therefore conclude that the value of the infimum is 0. It is, however, impossible to find a control u yielding this value (why?). Notice how the convexity of the integrand F with respect to the control u fails.

4. ANOTHER FORMAT

The cost functional of an optimal control problem may incorporate another term depending on the final state of the system. In general, we will have an objective of the form

$$I(u) = \int_0^T F(t, x(t), u(t)) \, dt + \phi(x(T)),$$

where the final state $x(T)$ is free, the function ϕ is assumed to be differentiable, and we have the typical state equation completed with initial conditions

$$x'(t) = f(t, x(t), u(t)), \quad x(0) = x_0.$$

This sort of, apparently more general, situation can indeed be reduced to our typical format by means of the trick

$$I(u) = \int_0^T \left[\frac{d}{dt} \phi(x(t)) + F(t, x(t), u(t)) \right] dt$$

$$= \int_0^T \left[\nabla \phi(x(t)) f(t, x(t), u(t)) + F(t, x(t), u(t)) \right] dt.$$

Therefore, it is in fact one of our typical examples for the new integrand

$$\tilde{F}(t, x, u) = \nabla \phi(x) f(t, x, u) + F(t, x, u).$$

The Hamiltonian for this new problem would be

$$\tilde{H}(t, x, u, p) = \nabla \phi(x) f(t, x, u) + F(t, x, u) + p f(t, x, u).$$

We notice that optimality conditions are the same compared with those for the functional I without the term $\phi(x(T))$, but with multiplier

$$\tilde{p}(t) = p(t) + \nabla \phi(x(t)).$$

The only change relates to the transversality condition, which now reads

$$\tilde{p}(T) = \nabla \phi(x(T)) \quad (p(T) = 0).$$

Hence, the computation of optimal solutions of such an optimal control problem proceeds in the same way, ignoring the contribution $\phi(x(T))$, which enters in writing the transversality conditions at T, as pointed out.

Example 6.15   *Imagine that we would like to find the optimal control to minimize the cost given by*

$$I(u) = \frac{1}{2} x(T)^2 + \int_0^T u(t)^2 \, dt$$

with state equation and initial conditions

$$x'(t) = -x(t) + u(t), \quad x(0) = x_0.$$

Since the hypotheses for the sufficiency of optimality conditions hold, we can find them by applying Pongryagin's principle. As we have explained before, those conditions are the same as those for

$$\int_0^T u(t)^2 \, dt$$

but with the corresponding transversality condition $p(T) = x(T)$. Thus, for the Hamiltonian

$$H(t, x, u, p) = u^2 + p(u - x),$$

we must solve the system

$$p' = p, \quad 2u + p = 0, \quad x' = u - x,$$
$$x(0) = x_0, \quad p(T) = x(T).$$

After a few computations, we have to solve the problem

$$x'(t) = -x(t) - \frac{1}{2} x(T) e^{t-T}, \quad x(0) = x_0.$$

The solution is

$$x(t) = \frac{x_0}{5 - e^{-2T}} (5 e^{-t} - e^{t-2T}),$$

and the optimal control

$$u(t) = -\frac{2 x_0 e^{-T}}{5 - e^{-2T}} e^{t-T}.$$

Under restrictions on the size of the control $u(t) \in K$, the methodology is similar.

5.   SOME COMMENTS ON THE NUMERICAL APPROXIMATION

Optimal control problems are so important in engineering that the simulation and numerical approximation of them has received considerable attention, surely more than variational problems. Several successful strategies have been analyzed and implemented (for instance, two-point boundary methods and schemes based on optimality conditions). This is again beyond the scope and the aim of this text. Our goal in this section is to provide a few basic ideas directly based on discretization and optimization (not on optimality conditions) that can help in understanding and appreciating the role and difficulties of discretization in optimal control problems. A good reference is [32], in which a systematic approach to approximation in optimization problems, including optimal control, is developed in a rather complete and exhaustive way.

The numerical approximation of optimal control problems could be analyzed as we did with variational problems, namely, by dividing the time interval into several subintervals and, assuming that admissible controls are constant on each of those subintervals, finding the best such feasible control. When the partition of the interval is sufficiently fine, we expect to calculate a fairly good approximation to the true optimal control of our problem. This, in principle, can be set up in this way, and optimal solutions can be approximated by utilizing numerical algorithms for nonlinear programming. Consider the following situation.

Example 6.16 *We will perform a numerical approximation of the following optimal control problem:*

$$\text{Minimize} \quad I(u) = \int_0^1 u(x)^2 \, dx$$

subject to

$$x''(t) = u(t), \quad u(t) \in K, \quad t \in (0, 1),$$
$$x(0) = x'(0) = 0, \quad x(1) = 1, \quad x'(1) = 0.$$

The physical interpretation is clear. We would like to minimize fuel expenditure, measured by the integral of the square of the control, when a mobile object is to travel a distance 1 in a straight line and finish with vanishing velocity. The brake/accelerator must belong to a preassigned set K.

Let

$$u = (u_j), \quad j = 1, \ldots, n,$$

be our independent variable, where u_j is the value of the control on the interval

$$\left(\frac{(j-1)}{n}, \frac{j}{n} \right).$$

By recursion, we can solve the state equation on each of those subintervals. Indeed, if

$$x'\left((j-1)/n\right) = a_{j-1}, \quad x\left((j-1)/n\right) = b_{j-1},$$

are the final values obtained for x' and x in solving the state equation in the interval $((j-2)/n, (j-1)/n)$, then we should solve

$$x''(t) = u_j, \quad t \in ((j-1)/n, j/n),$$
$$x'\left((j-1)/n\right) = a_{j-1}, \quad x\left((j-1)/n\right) = b_{j-1},$$

and put

$$a_j = x'\left(j/n\right), \quad b_j = x\left(j/n\right).$$

In this simplified situation, all computations involved can be done explicitly, and we obtain

$$a_j = a_{j-1} + \frac{u_j}{n}, \quad b_j = b_{j-1} + a_{j-1}\frac{1}{n} + \frac{u_j}{2n^2}.$$

The initial conditions imply $a_0 = b_0 = 0$. By using these recursive formulas appropriately, it is not hard to check that

$$a_j = \frac{1}{n}\sum_{k=1}^{j} u_k, \quad b_j = \frac{1}{2n^2}\sum_{k=1}^{j}(2j - 2k + 1)u_k.$$

The final constraints $x(1) = 1$, $x'(1) = 0$ translate into

$$\frac{1}{2n^2}\sum_{k=1}^{n}(2n - 2k + 1)u_k = 1,$$

$$\frac{1}{n}\sum_{k=1}^{n}u_k = 0.$$

We can even write these two constraints as

$$\sum_{k=1}^{n} ku_k + n^2 = 0, \quad \sum_{k=1}^{n} u_k = 0.$$

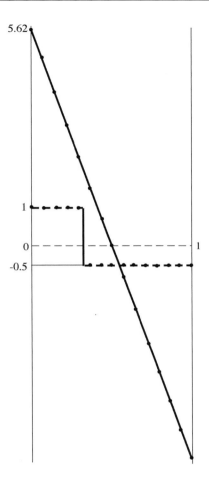

Figure 6.7. Approximation of an optimal control problem.

On the other hand, the cost functional is simply

$$I(u) = \sum_{k=1}^{n} u_k^2 .$$

Finally, we are faced with the problem

$$\text{Minimize} \quad \sum_{k=1}^{n} u_k^2$$

subject to

$$\sum_{k=1}^{n} k u_k + n^2 = 0, \quad \sum_{k=1}^{n} u_k = 0, \quad u_k \in K.$$

We have implemented two situations for different choices of the set K. The first one corresponds to no real constraint, so that $K = \mathbf{R}$. For the second one we have chosen $K = [-1/2, 1]$. One of the numerical algorithms of Chapter 4 can be used to find these discrete approximated solutions. Figure 6.7 shows the approximation for both cases.

Example 6.17 *We explain how the numerical approximation of the optimal solution for Example 6.7 can be set up.*

Since $T > 0$ is unspecified (it is precisely the cost functional to be minimized), we must incorporate it as another variable. Let

$$u = (u_j), \quad j = 0, 1, \ldots, n,$$

be our independent variable, where we are taking $u_0 = T$, and u_j is the value of the control on the interval

$$\left(u_0 \frac{(j-1)}{n}, u_0 \frac{j}{n} \right).$$

By recursion, as in the preceding example, we can solve the state equation on each of those subintervals. Indeed, if

$$x' \left(u_0 (j-1)/n \right) = a_{j-1}, \quad x \left(u_0 (j-1)/n \right) = b_{j-1},$$

are the final values obtained for x' and x in solving the state equation in the interval $(u_0(j-2)/n, u_0(j-1)/n)$, then we ought to solve

$$x''(t) = u_j, \quad t \in \left(u_0(j-1)/n, u_0 j/n \right),$$
$$x' \left(u_0(j-1)/n \right) = a_{j-1}, \quad x \left(u_0(j-1)/n \right) = b_{j-1},$$

and put

$$a_j = x'(u_0 j/n), \quad b_j = x(u_0 j/n).$$

As before, all computations involved can be done explicitly, and we obtain

$$a_j = a_{j-1} + \frac{u_0 u_j}{n}, \quad b_j = b_{j-1} + a_{j-1}\frac{u_0}{n} + \frac{u_0^2 u_j}{2n^2}.$$

The initial conditions imply $a_0 = b_0 = 0$. Again, it is not hard to check that

$$a_j = \frac{u_0}{n}\sum_{k=1}^{j} u_k, \quad b_j = \frac{u_0^2}{2n^2}\sum_{k=1}^{j}(2j - 2k + 1)u_k.$$

The final constraints $x(T) = \alpha$, $x'(T) = 0$ translate into

$$\frac{u_0^2}{2n^2}\sum_{k=1}^{n}(2n - 2k + 1)u_k = \alpha, \quad \frac{u_0}{n}\sum_{k=1}^{n} u_k = 0.$$

Or even further,

$$u_0^2 \sum_{k=1}^{n} k u_k + \alpha n^2 = 0, \quad \sum_{k=1}^{n} u_k = 0.$$

On the other hand, the cost functional is simply

$$I(u) = u_0.$$

Finally, we are faced with the problem

$$\text{Minimize} \quad u_0$$

subject to

$$u_0^2 \sum_{k=1}^{n} k u_k + \alpha n^2 = 0,$$

$$\sum_{k=1}^{n} u_k = 0, \quad -b \le u_k \le a, \quad u_0 \ge 0.$$

Notice how in optimal control problems the discretized version of the underlying optimization problem requires solving a differential equation. In the examples examined before, this has been done explicitly. In many situations, even simple situations, we cannot expect to be able to do that, so that either we have to incorporate a solver for differential equations as part of the (numerical) definition of cost functionals and/or constraints, or else we would have to see how to use the information coming from optimality conditions. This last possibility would be the subject of a more specialized text. We invite our readers to work out the details of the numerical approximation in another, more involved, but standard example: the optimal control of a harmonic oscillator (Example 6.9, Exercise 18). This time a numerical integrator (Euler integrator) to approximate the state equation over the subintervals where the control is constant should be utilized.

6.   EXERCISES

1. A system is governed by the state equation $x' + ax = u$, where a is a constant, $x = x(t)$ is the state, and $u = u(t)$ is the control. If $x(0) = 0$ and $x(T) = C$, determine the optimal control that minimizes the cost

$$I(u) = \int_0^T \left[(C - x)^2 + u^2 \right] dt.$$

Here C is also a constant.

2. A certain system obeying the state equations

$$x_1' = x_2, \quad x_2' = -x_1 + u,$$

starts at $x_1(0) = x_2(0) = 1$. Find the optimal control to have the system in the same state after one unit of time $x_1(1) = x_2(1) = 1$ if the cost is

$$I(u) = \frac{1}{2} \int_0^1 u^2(t) \, dt.$$

3. The equations

$$x_1' = x_2, \quad x_2' = -x_2 + u,$$

characterize the behavior of a system. If we consider a cost functional of the type

$$I(u) = \int_0^\infty (x_1^2 + \frac{16}{3} u^2) \, dt,$$

find the optimal control if

$$x_1(0) = a, x_2(0) = b, \quad x_1(t), x_2(t) \to 0 \text{ as } t \to \infty.$$

4. A system is governed by the equation

$$x'' + x' + x = u, \quad x(0) = c_0, x'(0) = c_1.$$

The control u is restricted by $|u| \leq 1$. Study the optimal control leading the system to rest in minimum time.

5. A rocket travels upward over ground under a constant gravitational force and negligible aerodynamical effects. The thrust of the engine acts vertically downward. The equations are

$$h' = v, \quad v' = -g + \frac{c\beta}{m}, \quad m' = -\beta,$$

where h is the height measured with respect to the ground, v is the vertical velocity, m is the total mass of the rocket, c is a positive constant, and β is the control representing the flux of fuel subject to the constraint $0 \leq \beta \leq \bar{\beta}$. Assuming that at the beginning $t = 0$, we have $m = m_0 + m_\beta$, where m_β is the amount of fuel, $h = 0$, $v = 0$, determine the optimal control so as to achieve maximum height, supposing that we are allowed a single change in the control.

6. A company decides to hire a marketing firm with the objective of maximizing sales of a certain product. The relationship between the sales level and the publicity employed, measured through a function $A(t)$, is given by the law

$$s' = rA\left(1 - \frac{s}{M}\right) - \lambda s,$$

where M is the level of saturation of sales, λ is the decay rate of sales under no marketing, and r is a positive parameter. If the money spent on marketing is limited at every instant by $0 \leq A \leq \bar{A}$ and also globally by

$$B = \int_0^T A(t) \, dt,$$

for a given fixed B, determine the optimal strategy to maximize global sales

$$S = \int_0^T s(t) \, dt$$

over a fixed period of time.

7. A mobile object is controlled by the law

$$x'' + x' - 2x = u, \quad |u| \le 1.$$

If

$$x(0) = -\frac{1}{6}e^2 - \frac{1}{3}e^{-1} - \frac{1}{2}, \quad x'(0) = \frac{1}{3}e^2 - \frac{1}{3}e^{-1},$$

determine the optimal strategy U to lead this mobile object to rest,

$$x(T) = x'(T) = 0,$$

in minimum time. Assume $U(0) < 0$.

8. A system is governed by the equation

$$x'(t) = x(t) + u(t), \quad t \in (0, 10),$$

and starts at $x(0) = 100$. If the cost is given by

$$I(u) = \frac{1}{2}x(10)^2 + \int_0^{10} \left[3x(t)^2 + u(t)^2\right] dt,$$

determine the optimal control.

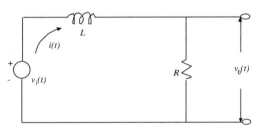

Figure 6.8. A circuit.

9. Consider the circuit of Figure 6.8. The initial current vanishes: $i(0) = 0$. A maximal voltage difference is desired,

$$v_0(T) = \int_0^T Ri'(t) \, dt$$

in the final instant T. The law of the circuit is

$$i'(t) = \frac{1}{L}v_i(t) - \frac{R}{L}i(t), \quad 0 \le v_i \le 1.$$

Determine the optimal strategy, and the maximum potential fall.

10. Two spaceships A and B travel in free space. In the initial instant, they are a distance c_0 apart, and B moves away from A at constant velocity c_1. A wants to reach B softly. The position of A is governed by

$$x'' = u, \quad x(0) = x'(0) = 0.$$

The energy consumed by A in this task is proportional to

$$I(u) = \int_0^T u(t)^2 \, dt.$$

Find the optimal strategy, having in mind that the link-up must be accomplished before a certain period of time T_1 ($T \le T_1$).

11. Try to solve the optimal control problem number 12 in Chapter 1.

12. A cup of coffee is initially at $100°$F, and we wish to decrease its temperature in minimum time to $0°$F by adding a fixed (unit) amount of milk. If $x(t)$ is the temperature of the mixture of coffee and milk in the cup, the law for the cooling of such a mixture is

$$x'(t) = -x(t) - 25u(t) - \frac{1}{4}u(t)x(t),$$

where $u(t)$ is the rate at which milk is added, and is restricted so that $0 \le u(t) \le 1$ and

$$\int_0^T u(t) \, dt = 1.$$

1. Argue why the optimal strategy must have the form

$$u(t) = \begin{cases} 0, & 0 \le t \le t_0, \\ 1, & t_0 \le t \le T, \end{cases}$$

for a certain $t_0 \ge 0$.

2. By bearing in mind the previous step, find the optimal strategy.

13. A certain plague is spoiling a crop. To fight against it, a predator is developed and introduced into the crop. Since the predator is also harmful to the crop as well as infertile, it is sought that both species be eliminated simultaneously as soon as possible. If $x_1(t)$ and $x_2(t)$ designate both species, respectively, and they start from $x_1(0) = 1/4$, $x_2(0) = 0$, determine the optimal control u and the minimum time if $-1 \leq u \leq 1$ and the state law is

$$x_1'(t) = x_1(t) - x_2(t), \quad x_2'(t) = -x_2(t) + u.$$

The control u represents the rate at which the predator is put or taken.

14. Sometimes, it is not known whether desired final states for a system can actually be reached under the constraints we have. In such cases, an optimal control problem like

$$\text{Minimize} \quad \frac{1}{2} |x(T) - x_T|^2$$

subject to

$$x' = f(t, x, u), \quad x(0) = x_0, \quad |u| \leq M,$$

may help in determining whether the desired final state x_T can actually be attained. This occurs when the optimal cost vanishes. For a linear state law with constant coefficients,

$$f(t, x, u) = ax + bu + c, \quad a, b, c \in \mathbf{R}, \quad a, b \neq 0,$$

decide whether there could be unreachable states x_T.

15. Pontryagin's maximum principle does not always provide optimal solutions. This is obviously so when there are no optimal solutions. Try to use Pontryagin's maximum principle for Exercise 13 of Chapter 1, and describe what kind of difficulties you encounter.

16. Examine the numerical approximation of Example 6.10 by using the ideas of Section 6.5.

17. Study the optimal control problem of leading the system governed by the state equation

$$x''(t) - x(t) = u(t), \quad -1 \leq u(t) \leq 1,$$

to rest from an initial arbitrary state in minimal time. Set up the scheme for the numerical approximation.

18. Explore the numerical approximation of the optimal control of a harmonic oscillator (Example 6.9) by using the ideas in Section 6.5.

References

These references are intended to provide a number of textbooks by which readers may deepen their knowledge and understanding of the different chapters treated in this text. It is not meant to be exhaustive. Some of the exercises and examples throughout the chapters have been inspired or literally taken from some of these references. In many instances, titles for textbooks are explicit enough so that readers may figure out their relationship with the different chapters of this book. Usually, most items in this set of references are more advanced (typically much more advanced) than the corresponding chapter. In other cases, references have been cited within the body of the text.

[1] Bazaraa M. S., Sherali, H. D., and Shetty C. M. *Nonlinear Programming, Theory and Algorithms*, J. Wiley and Sons, New York, 1993, second edition.
[2] Bertsekas, D. P., *Constrained Optimization and Lagrange Multiplier Methods*, Academic Press, Inc., New York, 1982.
[3] Bradley, S. P., Has, A. C., and Magnanti, T. L., *Applied Mathematical Programming*, Addison-Wesley, Reading MA, 1977.

[4] Burghes, D., and Graham, A., *Introduction to Control Theory Including Optimal Control*, Ellis Horwood, 1980.

[5] Burl, J. B., *Linear Optimal Control*, Addison-Wesley, 1999.

[6] Castillo, E., Conejo, A., Pedregal, P., García, R., and Alguacil, N. *Building and Solving Mathematical Programming Models in Engineering and Science*, J. Wiley and Sons, New York, 2002.

[7] Chvatal, V., *Linear Programming*, W. H. Freeman, New York, 1983.

[8] Clarke, F. H. *Optimization and Nonsmooth Analysis*, John Wiley and Sons, New York, 1983.

[9] Clarke, F. H., Ledyaev, Y. S., Stern, R. J., and Wolenski, P. R., *Nonsmooth Analysis and Control Theory*, Springer-Verlag, New York, 1998.

[10] Craven, B. D., *Control and Optimization*, Chapman and Hall, London, 1995.

[11] Conejo, A., *Técnicas de Optimización*, ETSI Industriales, UCLM, 2001.

[12] Dantzig, G. B., *Linear Programming and Extensions*, Princeton Univ. Press, 1993.

[13] Dennis, J. E., *Numerical Methods for Unconstrained Optimization*, SIAM, 1996.

[14] Ewing, G. M., *Calculus of Variations with Applications*, Dover, New York, 1985.

[15] Fiacco, A. V. and McCormick, G. P., *Nonlinear Programming: Sequential Unconstrained Minimization Techniques*, J. Wiley and Sons, New York, 1968.

[16] Fletcher, R. *Practical Methods of Optimization*, J. Wiley and Sons, 1990.

[17] Forsgren, A., Gill P. E., and Wright M. H., Interior Methods for Nonlinear Optimization, SIAM Rev., 44, 4, 525–597.

[18] Garfinkel, R. S. and Nemhauser, G. L., *Integer Programming*, J. Wiley and Sons, New York, 1984.

[19] Gelfand, I. M., and Fomin, S. V., *Calculus of Variations*, trans. ed. R. A. Silverman, Dover, Mineola, New York, 1991.

[20] Gill, P. E., Murray, W., and Wright, M. H., *Numerical Linear Algebra and Optimization*, Vol. I, Addison-Wesley, 1991.

[21] Gregory, J. and Lin, C. *Constrained Optimization in the Calculus of Variations and Optimal Control Theory*, Chapman and Hall, London, 1992.

[22] Hiriart-Urruty, J. B. and Lemaréchal, C. , *Convex Analysis and Minimization Algorithms*, Vols. I and II, Springer-Verlag, Berlin, 1996.

[23] Hocking, L. M. *Optimal Control: An Introduction to Theory with Applica-

tions, Oxford Univ. Press, 1997.

[24] Jahn, J., *Introduction to the Theory of Nonlinear Optimization*, Springer-Verlag, Berlin, 1996.

[25] Klamkin, S., ed., *Mathematical Modeling: Classroom Notes in Applied Mathematics*, SIAM, Philadelphia, 1987.

[26] Luenberger, D. G., *Optimization by Vector Space Methods*, J. Wiley and Sons, 1969.

[27] Luenberger, D. G., *Linear and Nonlinear Programming*, Addison-Wesley, 2nd ed., 1984.

[28] Mangasarian, O. L., *Nonlinear Programming*, SIAM, 1994.

[29] Mittlemann, H. D., and Spellucci, P. *Decision Tree for Optimization Software*, World Wide Web, http://plato.la.asu.edu/guide.html, 2003.

[30] Nocedal, A. J. and Wright, J. *Numerical Optimization*, Springer-Verlag, 1999.

[31] Pierre, D. A., *Optimization Theory with Applications*, Dover, New York, 1986.

[32] Polak, E., *Optimization: Algorithms and Consistent Approximations*, Springer, New York, 1997.

[33] Rao, S. S., *Engineering Optimization. Theory and Practice*, J. Wiley and Sons, New York, 1996, Third Edition.

[34] Rockafellar, R. T., *Convex Analysis*, Princeton Univ. Press, Princeton NJ, 1970.

[35] Strang, G., *Introduction to Applied Mathematics*, Wellesley-Cambridge Press, 1986.

[36] Troutman, J. L., *Variational Calculus and Optimal Control*, Springer, New York, 1996, Second Edition.

[37] Weinstock, R., *Calculus of Variations*, Dover, New York, 1974.

[38] Wright, S. J., *Primal-Dual Interior-Point Methods*, SIAM, 1997.

Index

Texts in Applied Mathematics

(continued from page ii)

Texts in Applied Mathematics

(continued after index)

Texts in Applied Mathematics 46